Political Geography

Political geography has produced some of the most radical and innovative ideas in human geography in the past twenty years. The meaning and significance of traditional political subdivisions, such as the state, have had to be fundamentally re-evaluated in the face of the globalisation of society and economy and this has forced political geographers to look for new ways of explaining the dynamics of the world system.

Political Geography provides a stimulating and concise introduction to the key themes of the subdiscipline, which moves beyond the study of the state to encompass the spatial consequences of power at all levels. It is divided into three sections:

- Process and Patterns considers how and why societies allocate and manage territory and why exclusive control over defined areas has become such an overwhelming preoccupation.
- Ideology and Geopolitical Visions looks beyond the state to evaluate critically the geopolitical strategies of successful expansionist powers that have enabled them to stamp their values and influence beyond their immediate political boundaries.
- Beyond the State extends the argument further to examine the impact that globalisation has had on political structures and on political relationships across the globe.

In the conclusion, the book asks whether the existing concepts and models used by political geographers are sufficiently robust to be able to contribute usefully to the analysis of the new world order in the twenty-first century. This student-friendly textbook is a valuable and accessible introduction to the broad themes of political geography.

Mark Blacksell is Emeritus Professor of Geography at the University of Plymouth.

Routledge Contemporary Human Geography Series

Series Editors:
David Bell, Manchester Metropolitan University
Stephen Wynn Williams, Staffordshire University

This series of texts offers stimulating introductions to the core subdisciplines of human geography. Building between 'traditional' approaches to subdisciplinary studies and contemporary treatments of these same issues, these concise introductions respond particularly to the new demands of modular courses. Uniformly designed, with a focus on student-friendly features, these books will form a coherent series which is up-to-date and reliable.

Existing Titles:

Urban Geography, second edition

Cultural Geography

Tourism Geography

Development Geography

Political Geography

Geographies of Globalization

Routledge Contemporary Human Geography

Political Geography

Mark Blacksell

LONDON AND NEW YORK

First published 2006
by Routledge
2 Park Square, Milton Park, Abingdon, Oxon OX14 4RN

Simultaneously published in the USA and Canada
by Routledge
270 Madison Avenue, New York, NY 10016

Routledge is an imprint of the Taylor & Francis Group

Typeset in Times and Franklin Gothic by
Keystroke, Jacaranda Lodge, Wolverhampton
Printed and bound in Great Britain by
TJ International Ltd, Padstow, Cornwall

British Library Cataloguing in Publication Data
A catalogue record for this book is available from the British Library

Library of Congress Cataloging in Publication Data
Blacksell, Mark 1942–
Political geography / Mark Blacksell.
p. cm. — (Routledge contemporary human geography series)
Includes bibliographical references and index.
1. Political geography. I. Title. II. Series.
JC319.B575 2005
320.1′2—dc22 2005006198

ISBN 0–415–24667–9 (hbk)
ISBN 0–415–24668–7 (pbk)

Contents

Figures

Boxes

Preface

It is just over a century since Sir Halford Mackinder published his seminal paper in the *Geographical Journal* on the geographical pivot of history, initiating the study of political geography as a serious academic endeavour in the English-speaking world. Political geography is now at the forefront of research and debate in human geography and firmly embedded in most advanced geographical curricula. Yet, it still does not always receive its due at more introductory levels and this book is an attempt to enthuse those early on in their geographical studies about political geography and its 'pivotal' role.

Undergraduate teaching is hugely rewarding, but success depends heavily on the availability of authoritative, up-to-date, and affordable textbooks that are easy to read and understand. This is especially true at a time when more and more people are taking up undergraduate studies, putting enormous pressure on teachers and lecturers, not to mention the resources of university and college departments. One would have hoped that there would be a premium in higher education on developing such books and other materials, but this is still too rarely the case, not least because of the academic prestige and money attached to high-quality research. The importance of such research is undeniable and teaching materials depend upon it for most of their vigour and renewal, but it does sometimes seem that the essential two-way link between teaching and research is being forgotten. I hope that in a small way this book will show what fun it can be to try to take a synoptic view of the vast and varied area of academic study that political geography has become.

As always, I am hugely indebted to others for enabling me to bring this book to fruition. Three anonymous readers provided many invaluable suggestions when the manuscript was at an early stage and I am indebted and very grateful for their help and advice. Neil Roberts, David Pinder, and all my colleagues at the School of Geography in the University of Plymouth have been a constant source of inspiration, challenge, and friendship and for that I thank them most warmly. I especially want to mention Brian Chalkley for his unswerving commitment to the promotion of quality teaching, which has not only been an exemplar for me, but I firmly believe for the discipline of geography as a whole. Brian Rogers, Tim Absalom, and Jamie Quinn, the cartographers in the School of Geography, have done marvels turning my rough sketches into finished maps and diagrams and the book would not have been completed without their unstinting professional efforts. Elsewhere, Helen Jones and Andrew Teed in the Geography Department at the University of Exeter provided invaluable help in finding and reproducing some of the historical materials. Uli Best encouraged me to delve into the critical literature, as well as alerting me to Albrecht Penk's foray into political geography. Thanks too to the following for giving me permission to reproduce material: Figure 2.1, the Royal Geographical Society with the Institute of British Geographers; Figure 2.3, Chris Walpole; Figure 4.4, the Open University Press/McGraw-Hill Publishing Company; Figure 6.1, Robin Hill; Figure 6.2, Thomson; Figures 6.3, 6.4, and 6.5, Pearson Education; Figure 10.1, Alistair Couper; Figure 10.2, the Controller of Her Majesty's Stationery Office and the UK Hydrographic Office.

Andrew Mould, the Commissioning Editor at Routledge, and his staff deserve a special thank you: for encouraging me to write the book; for their endless patience and good humour as they waited for the manuscript to be delivered; and for the efficiency with which they have seen it through to publication.

Finally, my love and thanks go to my family and friends, who have had to put up with so many obsessive hours devoted to political geography.

Mark Blacksell
University of Plymouth
February 2005

1 Placing political geography

Then lands were fairly portioned;
Then spoils were fairly sold:
The Romans were like brothers
In the brave days of old.

(Macaulay, *Lays of Ancient Rome*, 1842, Horatius, 31)

Political geography as a subdiscipline within contemporary human geography

One of the enduring fascinations of human society is the way in which the competing claims over the control and management of land and resources are played out. Unlike the mythical concord of Macaulay's ancient Rome, the reality is that individuals and interest groups of all kinds and at all levels are continually vying with each other to promote their own interests, thereby destabilising and changing the existing order and remaking the world in their own image. *Political geography, in the broadest sense, is the academic study of all these varied resource conflicts and the way in which they are resolved.* In other words, it is about the forces that go to shape the world we inhabit and how they play themselves out in the landscape across the globe. As such, it is relevant to everyone who is curious about the way we live and wants to understand better what might be happening all around them.

Political geography as a subdiscipline within contemporary human geography has seen its star rise and fall since it emerged as a distinct subdiscipline within geography at the beginning of the twentieth century, but it is now firmly established at the core. Political geographers have become increasingly important since the early 1970s in questioning the apparently self-evident orthodoxy that had typified much of economic and social geography for several decades previously. They have shown not only why generalised laws about human behaviour are by their very nature often flawed in practice, but also that individuals and minority groups matter geographically, in that they are invariably the initial drivers for change in the bigger picture, decisively influencing the terms of the debate and the way it evolves. One might say that people rather than peoples matter and that political geography has recently been at the forefront in incorporating that message into the geographical discourse (Painter, 1995).

The American geographer Edward Soja (1989) was amongst the most forceful of the critics, bemoaning the neglect of political economy in geographical research and writing and arguing for a more people-oriented approach. However, since the 1990s, such criticism has been increasingly hard to sustain as a growing band of radical geographers have challenged the traditional models and order, and begun to rewrite the foundations of human geography as a whole (see, for example, Harvey, 1989, 1996, 2000; Ó Tuathail and Dalby, 1998; Peet, 1998) (Box 1.1). David Harvey, more than any other individual, has been central to the debate, determinedly arguing the central importance of political economy for more than thirty years and gathering growing numbers of converts along the way. In *Spaces of Hope* (2000) he shows how the possibilities of globalisation and the image of a borderless world have been promoted by exploiting the possibilities of satellite imagery and other new technologies. He also shows how, somewhat ironically, this liberation has re-focused attention on the body and the needs of individuals. The big project now is how to bring these widely separated viewpoints together in a way that does disservice to neither: it is a challenge that political geography is well placed to accept and one that it cannot afford to ignore.

The nature, scope, and development of political geography

Political geography is an inherently dynamic subject and the nature and focus of the debates at its heart within modern geography have changed substantially over time. In what has been referred to, admittedly in a slightly different context, as 'a century of political geography' (Taylor, 1993: 1–7), it has been valued radically differently at different times, both within geography as a whole and outside. Inevitably, the different developmental phases do not fall neatly into discrete packages, clearly

Box 1.1

Modernism and postmodernism

Modernism is a cultural movement that rebelled against nineteenth-century Victorian values. Victorian culture emphasised nationalism and cultural superiority and placed humankind above and outside nature and the natural world. Modernists blamed the Victorians for evils such as slavery, racism, and imperialism, as well as later for the First World War. Modernists emphasised humanism over nationalism and argued for a much more liberal approach to cultural differences, which judged different cultures in their own terms, rather than in relation to European high culture. In their view, humankind was part of nature and there were multiple ways in which the world could be viewed, all equally valid. This in turn led them to reject decisively the Victorian distinction between the civilised and the savage, the driving force behind most of the great expeditions of imperial discovery in Africa, Asia, and the Americas.

Despite its more liberal view of the world, modernism was, nevertheless, characterised by a liking for grand theory and epistemology, valid knowledge or, more crudely, giving preference to an elitist view. Postmodernism, on the other hand, adopts a much more sceptical stance towards such grand claims and is highly suspicious of so-called fundamental laws and unchanging relationships that transcend such things as the constraints of time and space. Pluralism is, therefore, the main feature of postmodernism, along with an acceptance that any event, or situation, may be open to a potentially infinite variety of equally valid interpretations. The result is profound disagreement between postmodern scholars about the relative value of different interpretations, and a tendency to reject any kind of established orthodoxy. This has led many on the Left, including many geographers, to be extremely wary of accepting the postmodern movement as a real advance in the understanding of how societies work.

Reading

Harvey, 1989; Jenks, 1992.

separated in time, but Figure 1.1 illustrates in very general terms how the sequence of four phases fits together to form a logical progression.

In many ways political geography has always been a mirror of the times, very much reflecting current concerns, be it the lure of global empires, or a determination not to see individuality and the contribution of individuals swamped by higher-order priorities. The concentration on the current reality of people's lives has also led to marked variations of

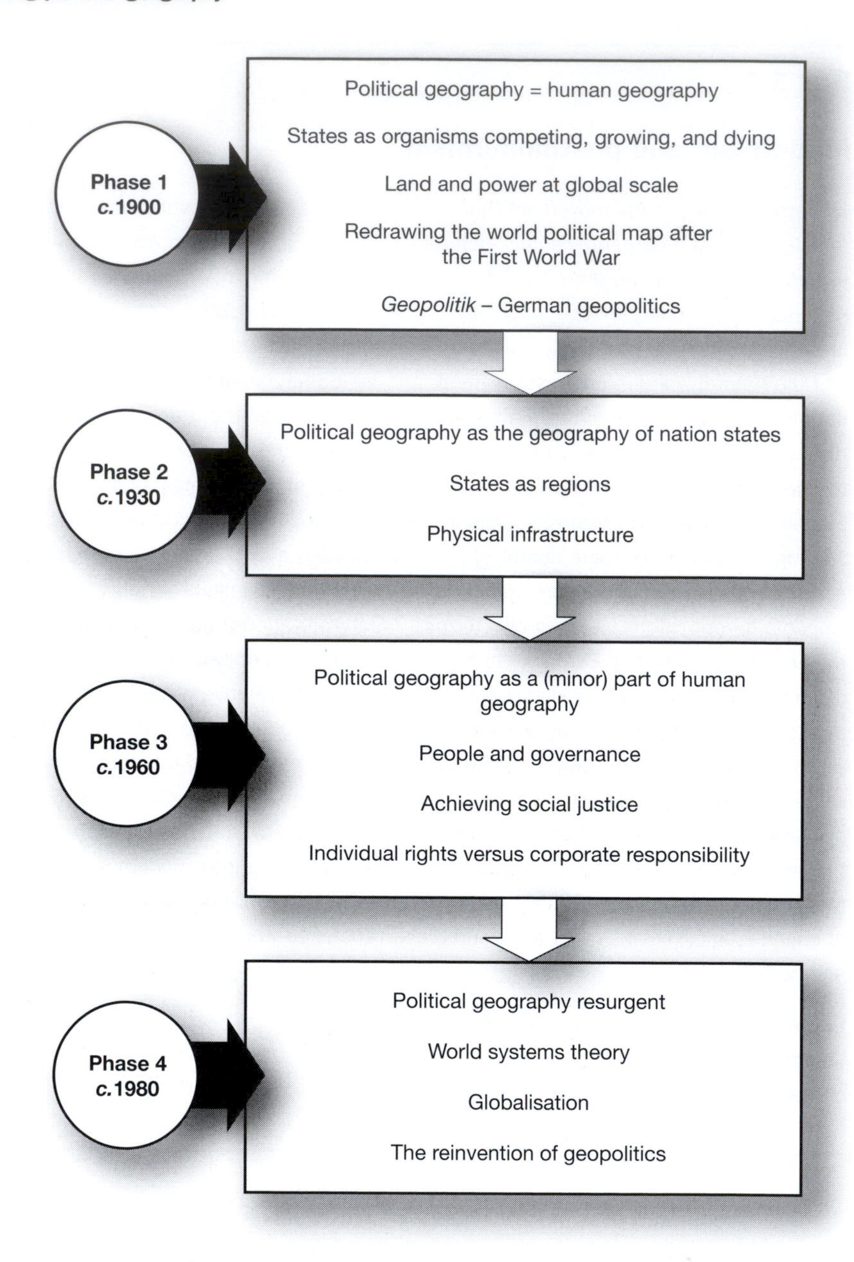

Figure 1.1 ***Major phases in the development of political geography***

emphasis between different societies and national academic traditions. Nevertheless, the roots of political geography (Figure 1.1, phase 1) lie firmly in the ferment of ideas about the nature of evolution emanating from Charles Darwin and his critics in the second half of the nineteenth century (see 'Environmental determination and the state' in Chapter 8). These ideas transformed scientific, and social scientific, thought at the time and have continued to influence the terms of much of the debate about the dynamic underpinnings of society ever since (Stoddart, 1966, 1981).

It was then, too, that modern geography began to be defined as a coherent academic discipline and political geography was seen as reflecting what would now be considered the whole range of human geography, rather than just one specialised part. The key person in this process was the German geographer, Friedrich Ratzel, whose *Politische Geographie*, first published in 1897, for the first time analysed systematically the dynamics of the relationship between human societies and the land on which they lived. He argued that states and peoples need the exclusive control and use of territory and that they will require ever increasing quantities as they develop and grow. The expansionist and competitive undertones of Ratzel's thesis are now viewed as highly controversial, potentially destabilising, and threatening to the prevailing political order, but their influence was nevertheless profound and long-lasting (Wanklyn, 1961; Cohen, 1964). It should also be remembered that at the end of the nineteenth century they were very much in tune with the general spirit of the age, with the hectic last throes of colonial expansion still in full swing and European states vying with each other in 'the scramble for Africa', which saw the whole of the continent except for Abyssinia – modern Ethiopia and Eritrea – under their colonial control. Any reservations about the morality of states seeking to expand their territory would have received short shrift at the time (see 'The spread of states' in Chapter 3).

On the contrary, Ratzel's ideas were enthusiastically built upon by geographers and used as the basis for interpreting the world order and how it might best be developed (Burghardt, 1969). In the forefront of this movement was the Englishman Sir Halford Mackinder, whose Heartland theory purported to explain the relationship between land and political power at a global scale, thereby demonstrating what he termed 'the geographical pivot of history' (Mackinder, 1904; Dodds and Sidaway, 2004). It is without doubt one of the most powerful ideas ever to be propounded by a geographer, explicitly influencing strategic thinking for more than a generation, notably through his best-selling book

Democratic Ideals and Reality: a study of the politics of reconstruction (1919). This global approach reached its apotheosis in the wake of the First World War, when Mackinder and a number of other geographers acted as official advisers in the negotiations that led to the Treaty of Versailles (1919), which began to put in place a new order, creating and recognising a radically revised political map in Europe and what had become the Soviet Union (see 'Geography and the world order' in Chapter 9). Indeed, the American geographer Isaiah Bowman, who was also part of the negotiating team, extended the practical proposals in the treaty into *The New World: problems in political geography* (1921), a global survey of states and other political jurisdictions in the post-First World War era and a justification for the essential rightness of their deliberations (Smith, 2003).

For political geography this can now be seen (with the invaluable gift of hindsight!) as a defining moment, initiating a shift of the academic focus away from a rather simplistic application of laws and theories developed in the natural sciences to social and political phenomena. Gradually states themselves became the centre of attention and the new orthodoxy was a drive to understand the essential components of a stable political entity (Figure 1.1, phase 2). What were the preconditions necessary for the huge variety of states, with all their different shapes, sizes, and geographies, to coexist without exercising predatory designs on each other's territory? It also had the effect of subsuming political geography into the wider debate about regionalism, which became the dominant discourse in geography as a whole in the second third of the twentieth century, with its focus on finding natural, self-sustaining regions at all levels of human activity, from the international and national to the local (Hartshorne, 1939).

Not that the taste for grand global designs disappeared without a significant last hurrah. As described above, Mackinder's Heartland theory still enjoyed considerable popularity and the frontier thesis of the American historian Frederick Jackson Turner continued to be the standard explanation for the inexorable spread of the United States across North America, as well as for the other continental-scale colonial expansions in Australia and southern Africa (Turner, 1894; Meinig, 1960; Kearns, 1984). However, in Germany the ideas of Ratzel and his followers enjoyed a true, if somewhat perverted, renaissance. Deprived of much of its former territory in Europe after its defeat in the First World War, as well as of all its overseas colonial territories in Africa and the Middle and Far East, Germany was eager to latch on to any argument for their reinstatement. A group of geographers worked enthusiastically to develop the pseudo-science of *Geopolitik*, which sought to justify a dynamic relationship between peoples, the land they occupied, and states, arguing that the latter must expect – and be allowed – to expand to accommodate legitimate aspirations for the exclusive political control of territory. It was by any standards a self-serving mission, which received more or less open support from the National Socialists of the Third Reich after 1933, a regime with the explicit aim of restoring Germany's former imperial greatness and the lands covering most of Central Europe that went with it (Bassin, 1987a).

The German *Geopolitik* movement did enormous damage to the credibility of political geography. The very idea that politics at the national level could and should be driven by inexorable geographical laws became an anathema and academic geographers retreated into approaches to human geography which all but excluded political considerations. For what was left of political geography as such, the state itself became the focus of attention, with the emphasis on issues such as the nature of frontiers and boundaries, and the centrifugal and centripetal forces working respectively to pull states apart or hold them together (Gottmann, 1952).

For all practical purposes the middle years of the twentieth century were an era of stagnation for political geography, devoid of any real theory and largely reliant on descriptive statements of the obvious. The leading edge of human geography was rooted in economic theory and spatial science, with the uncertainties associated with individual people and societies largely relegated to the comfort zone of regional geography. Once again this was very much in step with the tenor of the times, with the Cold War division of the globe into the Communist and free worlds accepted as a fact of life and a belief that conflicts within the free world could be resolved by rational argument, which all too often meant exclusively economic argument. Although a gross oversimplification of reality, it nevertheless successfully excluded the political from the greater part of the geographical debate and, in the process, banished consideration of the interests of minority groups of all kind to the margins.

As the apparent certainties and confidence of the Cold War years, and the global political world order that went with them, began to ebb away towards the end of the 1960s, there was a revival of interest in political geography. The focus, however, was less the state and competition between states at a world scale, than the conflicts and tensions over time and space between individuals and groups at all levels – international, national, regional, and local (Cox, 1973). Nor was the process only driven by a revival of interest in political geography itself. Economic and social geography, which for more than two decades had furnished the dominant paradigms in human geography as a whole, began to incorporate political conflict ever more prominently into their explanations of the workings of society; it was through this that political geography began to reassert its influence (Woods, 1998). Geographers who did not first and foremost see themselves as political geographers set in train the changes which led to a fundamental reassessment of the nature of political geography (Figure 1.1, phase 3). Pre-eminent amongst them was David Harvey whose book *Social Justice and the City* (1973) firmly demonstrated that life in cities varied, depending on who you were and where you lived, with rich and poor, men and women, and whites and people of colour inhabiting very different worlds. It showed that there was endemic and systematic discrimination against certain groups, leading to political conflict being an integral part of the urban experience. For instance, rather than urban planning being viewed as a politically neutral technical exercise, according to the new orthodoxy it came to be seen as an arena of conflict where the interests of different competing groups living in the city were played out. The fight to save the Jamaica Bay Wildlife Refuge from being sacrificed for an extension to the runway at Kennedy Airport in New York in the late 1980s is a classic example of such conflict. Another is the redevelopment of London's Dockland. The docks themselves had fallen into disuse and decay as they became too small and congested for the needs of larger, modern shipping, and eventually they were replaced by new inner urban office and commercial uses. However, the change was at the expense of existing local jobs and communities. The political conflict centred around whether the economic advantages of new investment should be allowed to outweigh the damage and disruption to the local population, especially as they had little chance of sharing directly in the benefits themselves, because of their lack of the necessary skills.

The stimulus for much of this new analysis originated from Marxist (Box 1.2) interpretations of social relationships, but it is now generally accepted that such conflict is fundamental to all types of society and that politics and political conflict are integral to understanding the dynamics of change. Indeed, political geography has been defined as 'the analysis of the systems of class/group conflict over time and space' (Dear, 1988, p. 270) and the new perspective has propelled it from the periphery to

Box 1.2

Karl Marx (1818–83)

Karl Marx was a philosopher, social scientist, historian, and revolutionary. He was also the most influential socialist thinker of the nineteenth century. He was born in Germany, but emigrated, first to Paris and later to England. London was his home from 1849 until his death in 1883 and it was here that he produced most of his influential writing. During his lifetime, discussion of his ideas was largely confined to a limited circle of European left-wing revolutionary thinkers, but after his death, his two best-known works, *The Communist Manifesto* (1848) and the three-volume *Capital* (1867, 1885–94), achieved worldwide popularity and influence. They formed the philosophical basis for the Communist revolutions that created the Soviet Union and the People's Republic of China, as well as those in many other states in Europe, Asia, and the Americas. The books made people understand much more clearly the importance of economic factors in shaping society, and their analysis of class structure and conflict transformed the study of history and social science, even though many of Marx's expectations about the future course of the revolutionary movement have never actually materialised.

In common with every other area of social science, Marx's influence on human geography has been profound, providing the dominant theoretical framework for analysis for most of the second half of the twentieth century. Marxist theory was particularly important in developing the understanding of the dynamics of urban areas, providing geographers, and others, with a cogent way of explaining how the modern city functions.

Reading

Castells, 1977; Gregory, 1978; Harvey, 1973; Lefebvre, 1992.

centre stage in the geographical literature. The majority of the major academic journals in human geography are now replete with studies of different kinds of local conflict and the ways in which they have been resolved; this is especially true of *Political Geography* (formerly *Political Geography Quarterly*), a journal founded in 1982 specifically as an outlet for the burgeoning research in political geography.

The renaissance in political geography does not stem solely from a greater appreciation of the complexities and tensions in local decision-making. Another penetrating insight, initially originating from Karl Marx in the middle of the nineteenth century, was the inevitable globalisation of the world economy. This too has been adopted and developed by political geographers into a theory explaining the nature of the relationship between states and peoples with markedly different levels of economic development (Taylor and Flint, 1999). World systems theory places the relationship between the developed and the developing worlds in an unequivocal perspective, illustrating the way in which industrial societies are so structured that they inexorably hold those that are less economically advanced in a dependent and subordinate position from which they have little chance of escape (Figure 1.1, phase 4).

The renewed attention being paid to the global scale has also reawakened the interest of political geographers in geopolitics. Freed of its predatory antecedents in interwar Germany, geopolitics is now a lively branch of political geography, analysing the way in which states relate to each other to form coherent interest groupings in an increasingly globalised and internationalised world (Ó Tuathail, 1998). Geopolitics as a political activity, of course, never went away, but for more than a generation after the Second World War, it was virtually ignored by political geographers, who were embarrassed by the uses to which it had been put in their name. It is now once again making a substantial contribution, not least in furthering understanding of the international free trade system that is at the heart of the modern global market economy (Agnew and Corbridge, 1995). The world today is dominated by international companies that dwarf many of the states seeking to encompass and control them. As a result, the relationship between private capital and the state has fundamentally changed from the traditional model where governments were largely able to dictate the context for international development. It would be facile to claim that the situation is now simply reversed, but the dominance of governments is being seriously challenged by the free movement of capital across national borders across the world, thereby creating a new sort of geopolitics that includes important actors other than the traditional states, such as transnational and multinational companies (see 'Transnational corporations' in Chapter 11).

The challenge for political geography is how to retain coherence in the face of demands that it interpret both an increasingly globalised political order and local populations that are ever more politicised. It is an exciting challenge and, as was pointed out above, one that is integral to understanding the dynamics of the modern world, where the revolution in communications has led to a fundamental restructuring of the relationship between time and space that is reshaping the structures of political power at all levels.

Learning outcomes – the logic and structure of the book

The aim of this book is to present a contemporary view of political geography and, to this end, it is divided into three major sections. Section A, Process and patterns, looks at the fundamental ideas that have shaped political geography and made the subject what it now is. The key to this is the *concept of human territoriality*, which reflects the apparently insatiable human desire to define exclusive territories over which people can exercise some degree of control (Chapter 2). To a very large degree *maps and the development of cartography* enabled them to do this, so that the evolution of maps and map-making are inextricably entwined with the development of political geography. As cartographic techniques became increasingly sophisticated, so the precision with which space has been divided up has become ever greater, to a point where maps are now normally the ultimate evidence where demarcation disputes over land and territory are concerned.

Once territories can be clearly and unequivocally defined, more formal political entities have to be devised. The most powerful of these is the state, traditionally one of the ultimate symbols of undisputed political control. Chapter 3 examines the ways in which the *idea of the state* have evolved and the phenomenal proliferation in the number of independent states during the nineteenth and twentieth centuries. It also touches on their limitations, particularly in the context of modern economic, social, and technological conditions. Chapter 4 considers internal workings of states and how they may be made to function as coherent entities, looking at the different mechanisms for allocating the *exercise of power and the distribution of resources*. It also examines how the internal structures may be made sufficiently flexible, so that they can accommodate the change necessary to ensure that states remain vibrant and relevant to the people they serve.

States are often presented as symbols of unity, coherence, and stability, encompassing within their frontiers a population united by a common allegiance to a political ideal. The reality, as Chapter 5 demonstrates, is that all states are made up of many different peoples, more or less loosely held together by a set of political ideals. *Differences in language, ethnicity, culture, gender, and religion* all combine to produce an, at times, volatile mix, which can sometimes erupt and threaten the very future of the state itself. At the extremes, such discontent will crystallise into calls for a new political order, which can lead to the creation of new states. *Irredentism*, where people feel closer to a neighbouring state than the one in which they actually reside, is a common and persistent phenomenon across the globe and is fundamental to explaining why the world political map is, and is always likely to be, in a constant state of flux.

To be successful, therefore, the internal structures of states must be sufficiently flexible to accommodate change and to incorporate new groups and new interests. In democracies this is catered for by freedom of association and the ability to form new political parties. Chapter 6 examines the process by which *informal pressure groups* develop and evolve into *established political parties* and the essential role that these play in the functioning of the political landscape. Central to the whole question of political legitimacy are elections, the system through which the changing will of the people is expressed. All states have elections, though the extent to which they are permitted to allow people to exercise real choice varies greatly. There is little choice in only being able to vote for a single candidate and all electoral systems curb to some degree the freedom of candidates to stand, even if only by requiring that they demonstrate minimal support and the ability to raise a financial deposit. Chapter 7 explores how the *geographical analysis of electoral results* has been used to uncover subtle differences in the internal political landscape of states and reveal pointers to the dynamics of change.

Section B, Ideology and geopolitical visions, shifts the focus from the individual state to *visions of global hegemony*. Powerful states have always sought to dominate others, either by military conquest, or by imposing their own particular versions of the truth. Empires have waxed and waned as a result, but in the past hundred years political geographers and academics from other disciplines have played an important part in providing justification for expansionist yearnings, which challenge the status quo (Chapter 8). The truth is that all states are ambitious and have some kind of world view, as Chapter 9 demonstrates, even if it is only limited to how they see themselves in relation to the rest of the world. Indeed, a mutual understanding of each other's respective world views is an essential ingredient in *maintaining global political stability*.

Many expansionist dreams and policies are based on a belief that land is not politically incorporated. Such justification was used extensively in the nineteenth century as European-dominated empires drew much of Africa, America, Asia, and Australasia within their orbit. In the second half of the twentieth century the equivalent, though much less bloody, conquests have been of the *oceans*. Until the late 1940s only a thin sliver of sea around the coast was considered part of the national territory of states, the rest, the high seas, were viewed as international waters and beyond national political control. This has now profoundly altered, with most states claiming ocean territory up to 200 hundred nautical miles from their coasts. As Chapter 10 shows, this has radically altered the global political map and generated new conflicts between states over access to resources in the sea and beneath the seabed.

The final section, Section C, Beyond the state, looks beyond the conventional political map at the limitations of the state idea for understanding the true dimensions of social and economic relations at a global scale. Chapter 11 considers *internationalisation* and the whole *concept of globalisation*, focusing on the questions of deep-rooted inequality that characterise the world and examining the extent to which dependency is a necessary precondition for the *functioning of the capitalist world economy*. Globalisation has certainly brought with it huge opportunities, not least in terms of *mass worldwide communication*, but it has also put many important issues beyond the control of states and other formal political institutions. *Environmental degradation and resource depletion* are characteristic of such issues and many companies and other economic entities are bigger and more powerful than the states that try to manage and control them. This then raises serious questions about the relevance of the existing political order for enabling people to maintain a modicum of control over their own lives.

The response has been a *proliferation of international agencies* in the course of the twentieth century (Chapter 12). Globally, through organisations like the United Nations, and regionally, through bodies such as the European Union (EU) and the North Atlantic Treaty Organisation (NATO), states have come together in ever closer union to try to meet the global challenge. However, the evidence is that such groupings are as vulnerable as states themselves and have always been subject to shifting allegiances and pressures for change.

Political geography and political geographers have undoubtedly contributed to the understanding of the processes of political change in the course of the twentieth century. Chapter 13, the final chapter in the book, looks forward to how they may now be able to contribute to our understanding of the *challenge of change* in the twenty-first century.

Key themes and further reading

This chapter conveys what is meant by political geography and how it has developed as a distinctive subdiscipline within geography. The differences between the four main phases in that development, as summarised in Figure 1.1, should now be clearly understood. The main concepts to be discussed in the rest of the book should also now be apparent.

A host of political geography textbooks have been written since Friedrich Ratzel first coined the term in 1897. In recent times, the most influential and accessible is *Political Geography: world-economy, nation-state and locality* by Peter Taylor, which was first published in 1985; the fourth edition is written jointly with Colin Flint (1999) and the book provides an essentially structuralist and Marxist approach to understanding the political geography of the world at a variety of different scales. An excellent introduction to the theory of political geography is provided by *An Introduction to Political Geography: space, place and politics* by Martin Jones *et al.* (2004), while *Political Geography* by M. I. Glassner and C. Fahrer (2003), now in its third edition, provides a compendious and very well-illustrated overview of the field. For a more advanced critique, *Making Political Geography* by John Agnew (2002) offers a most stimulating analysis, and *Politics, Geography and 'Political Geography'* by Joe Painter (1995) offers a wealth of insights into the links political geography has developed with other areas of social science, especially at the local level. To see political geography in action, *Social Justice and the City* by David Harvey (1973) is still a revelation, more than a generation after it was first published. More than any other book in recent times, it has changed the way in which geographers understand cities and the dynamics of settlements in general. All Harvey's books are thought-provoking, none more so than *Spaces of Hope* (2000). Taken together, they provide a fascinating insight into how human geography generally, including political geography, has evolved over a generation.

SECTION A

Process and patterns

2 Human territoriality, maps, and the division of space

I cannot help thinking that in discussions of this kind, a great deal of misapprehension arises from the popular use of maps on a small scale. As with such maps you are able to put a thumb on India and a finger on Russia, some persons at once think that the political situation is alarming and that India must be looked to. If the noble Lord would use a larger map – say one on the scale of the Ordnance Survey map of England – he would find that the distance between Russia and British India is not to be measured by the finger and the thumb, but by a rule.

(Lord Salisbury, House of Commons, 11 June 1877)

The challenge of space

In most modern societies space is there to be conquered, but the desire for mastery manifests itself in endless different ways, depending on the individual or group concerned and the particular nature of the perspective in question. Equally, as Lord Salisbury pointed out, there is always a great danger of misunderstanding and misinterpretation whenever questions relating to space are being discussed. One of the best illustrations of the potential for diversity is still the study by Wilkinson (1951) of the way in which the region of Macedonia, part of which is now an independent state of the same name in the central part of the Balkan peninsula, was defined by different ethnic and political groups from the middle of the nineteenth to the middle of the twentieth centuries. He analysed nearly one hundred different maps, all of which purported to depict Macedonia. None was the same and each illustrated a particular point of view, ranging from the different ethnic interests – Greek, Turkish, Slav, Serb, Albanian – to the territorial ambitions of Britain, Germany, and Russia (the Soviet Union). Wilkinson points out that the reasons for the diversity in part represent a willingness to misrepresent the facts in pursuit of selfish political ends, but also in part are evidence of simple ignorance, the changes in the political situation over time, not to mention a lack of consensus as to which criteria should be used to distinguish between people and areas.

In similar vein, Sinnhuber (1954) attempted to pin down what it was people were referring to when they wrote about Central Europe. As can be seen from Figure 2.1, he illustrated how much can be lost by assuming that translations are exact equivalents in their respective languages. Central Europe is loosely translated as *Mitteleuropa* in German and *l'europe centrale* in French, but each means something slightly different in spatial terms. When this is allied to variations in definition by writers within each language the scope for confusion is self-evident.

Nevertheless people do define themselves to a significant degree in terms of space, deriving their sense of identity from specified tracts of land, be it the nation state, house and home, or their religion (Tyner, 2004). Indeed, there are few activities in society that do not have a spatial referent. The fundamental drive towards what is known as territoriality is widely accepted as a key social process and must be clearly understood if the dynamics of society as a whole are to be properly interpreted (Sack, 1980 and 1986).

Formally, territoriality is defined as 'the strategy used by individuals, groups and organisations to exercise power over a portion of space and its contents' (Agnew, 2000, p. 823) and, as a strategy it contains a number of components, three of which stand out. First of all, territoriality is a form of classification by area, which both includes and excludes. In the case of a state, the majority of those living within its borders are citizens and the majority of those outside are not, although within both groups there will be a certain fuzziness at the edges. Some people living within the borders, such as *Gastarbeiter* (guest workers) in Germany, will not have full rights of citizenship, whilst others outside, like diplomats working in embassies overseas, will have.

Second, territoriality must be communicated, either physically on the ground, or through some form of easily decipherable graphical representation on a map or plan. One has only to look at the average house to see how important communicating territoriality is. The urge to proclaim your exclusive right of access and to distinguish your own property from that of everybody else is almost overwhelming and in the UK results in the bewildering profusion of idiosyncratic gardens for which the nation is famous.

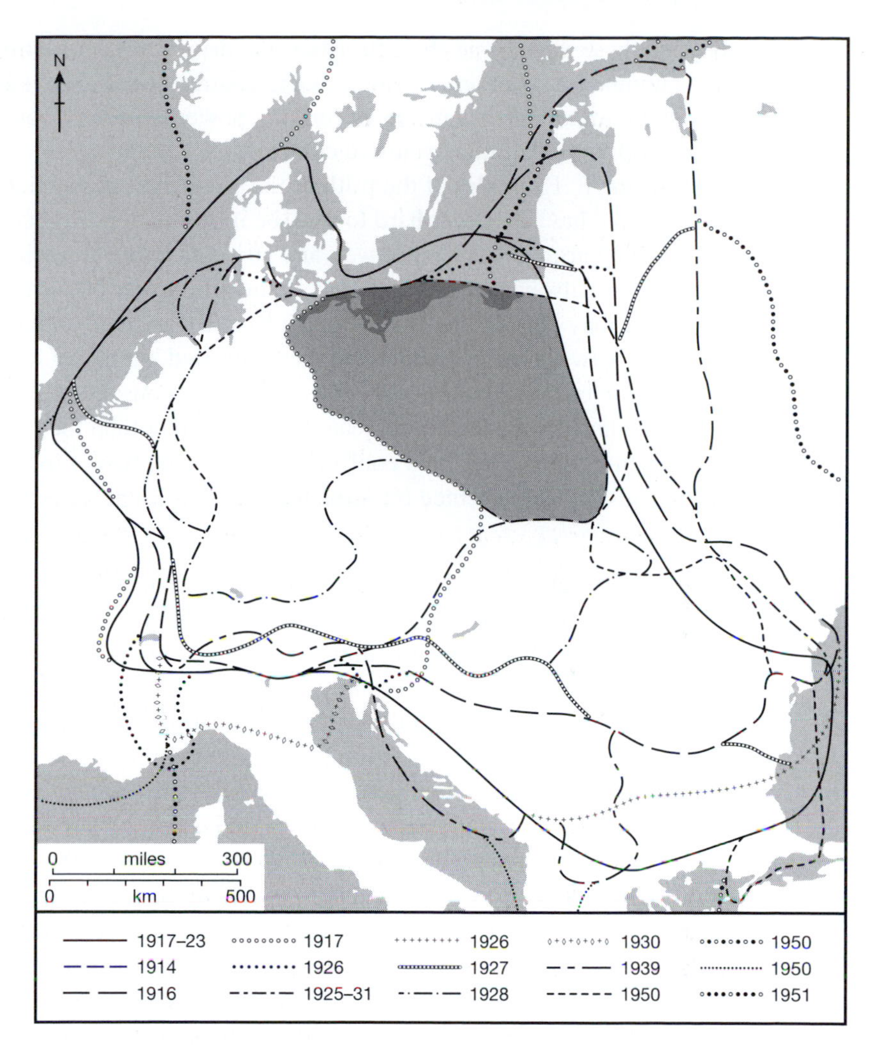

Figure 2.1 ***Mid-twentieth-century definitions of Central Europe,* Mitteleuropa, *and* l'europe centrale**

Note: The shaded land area shows where the various definitions between 1914 and 1951 overlap.
Source: After K. Sinnhuber (1954) 'Central Europe – Mitteleuropa – l'europe centrale: an analysis of a geographical term', *Transactions and Papers, Institute of British Geographers* 20(1): 19. Reproduced with permission of the Royal Geographical Society with the Institute of British Geographers.

Finally, the maintenance of territoriality demands enforcement, either through the physical presence of the police and the military, or through threats of recourse to the law and direct action should claims be ignored. 'Trespassers will be prosecuted' is a common message in the countryside and it is not infrequently backed up by that secondary warning 'Beware of the dog', just to ram home the consequences of not complying.

Beyond these three basic characteristics, the drive for territoriality also has a number of other less obvious characteristics. It has always been extremely important as a means of reifying power in society, for successfully proclaiming exclusive control goes a long way to legitimising it. Examples of the truth of the old adage 'possession is nine-tenths of the law' are never hard to find, be it in squatters' rights or land occupied by armies as an act of war, and attempts to reclaim what has been lost are always fraught with uncertainty, no matter how strong the legal case may be. Territoriality also displaces the fundamental balance of power between the controller and the controlled, replacing it with a relationship enshrined in law. This has the further consequence of depersonalising what has been created, so that rather than the balance of power being represented by a confrontation between two or more individuals, it is represented by something such as a statute, a by-law, or a contract.

Territoriality self-evidently leads to space being divided into containers within which people live and activities occur, indeed the spaces created actually mould what goes on within them. There are many situations where spaces are deemed to be empty, not because they are actually empty, but because they lack the particular functions thought to be appropriate to them. A classic example of such culturally crafted emptiness are the brownfield sites in towns and cities, which are seen as underutilised land, ripe for redevelopment and an alternative to the further outward spread of urban sprawl. In fact, of course, they are as 'full' as the 'developed' sites that surround them, but not with the artefacts (or in some cases the people) demanded by society as a whole (Sibley, 1995). Conversely, the space allocated is very often insufficient for all the people and activities that need to be contained and under these circumstances territoriality becomes inherently expansionist, seeking by whatever means to annex more space to fulfil its perceived destiny.

Territoriality is, therefore, an artificial political construct that seeks to subdivide space. It is inherently political, dynamic and, by its very nature, controversial. Without the remorseless drive to partition and make exclusive allocations of space the world as we know it would not exist, but there is nothing remotely natural or absolute about what has been created. It is inherently contested and ephemeral, but for all that represents the reality of most people's daily experience.

Formalising control

Natural and cultural landscapes rarely subdivide neatly along the same lines and attempts to use natural features as political boundaries have hindered as much as they have helped. Even apparently clear physical boundaries, as between land and sea, usually turn out to be unhelpful in practice, because of the way they deny the importance of human interaction between different biomes in favour of interactions within them (Box 2.1). Rivers probably represent the most obvious example of the dilemma. They are classic zones of contact and interaction and, as a result, are often the first linear features to be formally represented on maps and other media. They therefore appear as natural dividing lines, whereas in reality they form natural meeting places and zones of interaction. Indeed, there has been a strong counter lobby for using river catchments and basins as political and management entities, one of the most notable example being the Tennessee Valley Authority in the USA. However, although much quoted and celebrated since its inception in the 1930s, it never lived up to expectations in political terms and attempts to copy it, both in America and elsewhere in the world, have met with even more limited success (McKinley, 1950). The message is clear: the natural and the political do not necessarily coincide and attempts to make them do so are likely to end in failure, not least because of the fundamental differences in the dynamics of the physical and cultural environments.

Political legitimacy and control crucially depend on three things: symbols, systems, and agency and all must operate effectively if political areas at all levels are to be legitimated and enjoy any semblance of stability and permanence.

Symbols

Concrete symbols of control are essential; they remind people as to who is in charge and continually reinforce the message when evidence on the ground is often hard to find. Examples of such symbols are everywhere in a plethora of signage: at national frontiers you are threatened or welcomed, depending on the state of relations between the two adjoining countries, and within national jurisdictions local boundaries of all kinds are replete with warm words, ushering you in and waving you out (Figure 2.2). Nor is this a recent phenomenon, even though the scale has grown immeasurably in line with the increase in travel. The tradition of

Box 2.1

Tennessee Valley Authority (TVA)

The TVA was one of the innovative solutions put forward by President Franklin D. Roosevelt as part of his New Deal initiative to try to lift the USA out of the economic depression of the late 1920s and early 1930s. He asked the Congress to create 'a corporation clothed with the power of government but possessed of the flexibility and initiative of a private enterprise'. On 18 May 1933 the TVA Act was passed, setting up an authority encompassing the whole of the Tennessee river basin and with powers to control all the economic activities within this huge area.

From the outset the TVA adopted an integrated resource management approach to its task. Each issue, whether it was power production, navigation, flood control, malaria prevention, reforestation, or erosion control, was studied in its broadest context and each issue was weighed in relation to the others. Even by the standards of the Depression, the Tennessee Valley was in a desperate state in 1933. Much of the land had been farmed too hard for too long, eroding and depleting the soil. Crop yields had fallen along with farm incomes. The best timber had been felled. The TVA developed fertilisers, taught farmers how to improve crop yields, and helped replant forests, control forest fires, and improve habitat for wildlife and fish. The most dramatic change came as a result of the huge amount of electricity generated by the TVA dams on the river system. Electric lights and modern appliances made life easier and farms more productive. Electricity also attracted industries in to the region, providing much needed new employment.

During the Second World War, the USA needed aluminium to build bombs and aeroplanes, and aluminium plants required electricity. To provide power for such critical wartime industries, the TVA began one of the largest hydroelectric construction programmes ever undertaken in the USA. Early in 1942, when the effort reached its peak, 12 hydroelectric projects and a steam plant were under construction at the same time, and employment in design and construction reached a total of 28,000.

By the end of the war, the TVA had completed a 1,050-km navigation channel along the length of the Tennessee River and had become the nation's largest electricity supplier. Even so, the demand for electricity was outstripping its capacity to produce power from its dams. Political opposition prevented the TVA from obtaining additional federal funds to build coal-fired power stations, so it sought the authority to issue shares. The Congress passed the necessary legislation in 1959 to make the TVA power system self-financing, and henceforth it had to pay its own way.

The 1960s were years of unprecedented economic growth in the Tennessee Valley.

The farms and forests were in better condition than for generations. The cost of electricity was amongst the lowest in the country and it stayed low as the TVA bought larger and more efficient generating units into operation. In the expectation that the demand for electricity would continue to grow, the TVA began to build nuclear plants as a new source of cheap power. However, the economic downturn sparked by the international oil crisis in 1973, and the rapid increase in fuel costs later in the decade, brought significant changes to the economy of the Tennessee Valley. The average cost of electricity increased fivefold from the early 1970s to the early 1980s and, with the demand for energy falling and construction costs rising, the TVA was forced to cancel several nuclear plants. The TVA was forced to become more competitive and by the late 1980s it had stabilised power prices and begun to restructure the operation of its generating capacity. It cut its operating costs by $800 million a year, reduced its workforce by more than half, increased the generating capacity of its plants, stopped building nuclear plants altogether, and produced a long-term plan to meet the energy need of the Tennessee Valley as far ahead as 2020. It also implemented an extensive programme to reduce all forms of pollution from its plants. It is now third lowest of all the 25 major power generating companies in the USA, but the TVA still struggles to remain competitive and meet the obligations to integrated resources management that led to it being created in the first place.

regularly beating the parish bounds in England goes back to medieval times, and involves the whole community periodically walking around the perimeter of the parish checking that the boundary stones and other markers are still in place and, crucially, that they have not been surreptitiously moved by neighbouring parishioners eager to garner a little more land for their own parish (Figure 2.3).

Other kinds of symbol further bolster a sense of cohesion. Postage stamps remind people of the role of the state in promoting and controlling the means of communication between citizens and with the world beyond national borders (Stamp, 1966). It is fascinating that one of the most striking features of the current IT revolution is the dominance of private multinational companies, which operate largely outside national influence and control (see 'Transnational corporations' in Chapter 11). The change represents, in symbolic terms at least, a significant ceding of power by the state. In the economic sphere, coinage and bank notes send similar messages as to who is in control, though here too the emergence of a few dominant world currencies reflects an important aspect of changing political realities (Unwin and Hewitt, 2001). In the early part of the twentieth century, the pound sterling and

Figure 2.2 ***Road sign in English and Cornish at the border between Cornwall and Devon***

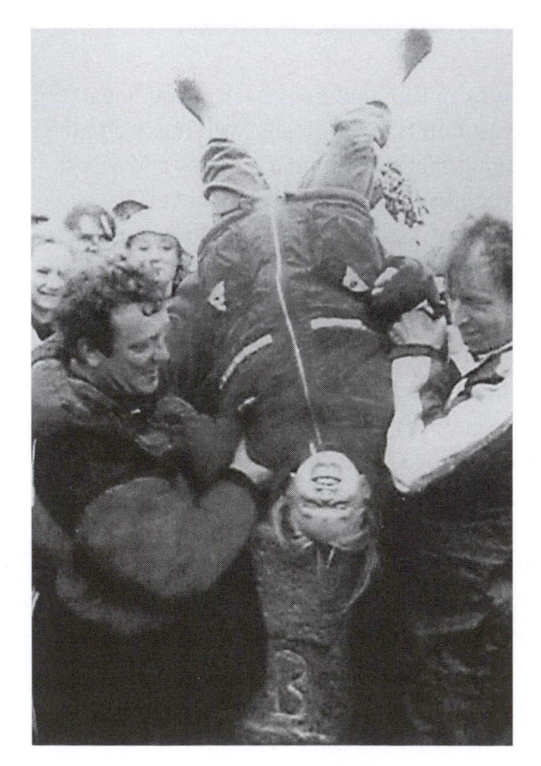

Figure 2.3 ***Beating the parish bounds in Belstone. Every seven years the ritual of walking the boundary of the parish is observed and young children are upended and have their heads (lightly) bumped on the main boundary marker stones to ensure that they are firmly imprinted with the territorial ethos of the parish***

Source: Chris Walpole, with permission.

the French franc were international trading currencies, with spheres of influence stretching way beyond their national borders and encompassing the whole of their respective colonial empires. In the second half of the twentieth century, the US dollar achieved an even more extensive sway and one that it continues to enjoy in much of the world. At the turn of the twenty-first century, however, the position of the dollar in Europe is being challenged by the euro, a common currency for 12 of the 25 members of the European Union. Leaving aside the economic arguments for and against adopting the euro as a national currency, it is an interesting case, because whether or not the currency is adopted sends a strong message about the willingness in the countries concerned to modify their political priorities in line with changing economic imperatives.

Some of the most potent symbols of national control stem from transportation and transportation networks. In most countries roads and road-building are public responsibilities and at least from the time of the Roman Empire some two thousand years ago roads have been viewed as a way of binding land together into a unified territory. In recent times, the motorway system in Germany was begun in the 1930s under the Third Reich with the explicit aim of making Berlin more accessible to Bavaria and the south and west respectively. Subsequently, most other European countries have followed suit and now also have extensive national motorway networks. It is also interesting that one of the first major initiatives to encourage European integration was a project to link these national motorway networks and the so-called E routes now connect most of the major cities in Europe in much the same way that the interstate road system does in North America. As with the example of the euro discussed above, such integration also has the political effect of shifting the focus away from the national to, in this case, the European level.

Similar national involvement and support is to be found with railways, airlines, and the other major carriers of goods and people. Even where the operators are private companies or the network privatised, there is usually still a strong sense of identity with the prevailing national or regional political image. Ships on the high seas fly the flag of the country where they are registered; and airlines are invariably closely identified with their country of origin, even where they are not formally national carriers. It all serves to bind territory and allegiance closely together in the popular imagination and thus reinforce the sense of commitment to an arbitrary and ultimately ephemeral political reality.

Systems

Symbols on their own are, of course, too superficial to be able to provide the necessary coherence for political survival. They must be backed up by formal systems which codify and explain the nature of the state, or other political area, and how it is supposed to function. For most states the critical document is the constitution, which provides a blueprint embodying everything for which it stands (Duchacek, 1973). The constitution is then elaborated and supported by a range of other entities, such as legal codes and political institutions, all of which derive their authority from the constitution itself. Amongst other things, these will determine how contracts are enforced and disputes settled, between both individuals and the state and between two or more individuals. This then is the key to the orderly management of territory, enabling private and public property rights to be guaranteed and providing for orderly transfers when the occasion demands (Waldron, 1990).

Historically, formal written constitutions are a relatively recent innovation, the first modern example being the American Constitution, which was adopted by thirteen states along the American eastern seaboard and came into force in 1788. There are still many states that do not have a written constitution, depending instead for their legitimacy on the collected customs, government, and law of the land, rather than on a single written document. Nevertheless, the UK, which is probably the prime example of a constitutional state without a written constitution, certainly operates a system of government that allows its territory to be managed effectively within an established and agreed framework of behaviour, as do other such states.

Agency

Finally, there needs to be appropriate agency to ensure that what is written down and supposed to happen is actually enforced. To this end, there must be military and police to try to guarantee compliance by upholding the law, though it is important to stress that this on its own is not sufficient. One of the purposes of formal education, the bulk of which is provided by governments, either completely free or with substantial public subsidy, is to instil a sense of identity and shared purpose. While it would be a crude exaggeration to imply that the main aim of educational provision is national indoctrination, to a degree it is part of the educational agenda and explains why political control over education is so hotly contested.

Political maps and cartography

Cartography is the classic artifice for representing and recording the division of space and maps, one of the tangible results of cartographic skills, have always been at the heart of geography. Their capacity to store and reproduce information about the world at innumerable different scales and to relate the results accurately one to the other has also made maps inherently and intensely political. As Harley has so ably and effectively demonstrated, maps provide knowledge, confer power, and historically have made map-makers sought after people by political elites; not for nothing was cartography known in medieval times as 'the science of princes' (Harley, 1988).

The power of maps and their innate ability to create a new reality for those who use them has long been recognised and valued. The University of Pennsylvania in the USA has an example of a map from Mesopotamia that is three thousand years old, with cuneiform characters incised on a clay tablet. All major societies since have made use of maps to depict boundaries and curtilages at every conceivable scale, thereby legitimating politically the division of the territory they control (Wood, 1993). The power they embody also means that access to maps has traditionally been jealously guarded: they were the preserve of religious elites in ancient Egypt and Christian medieval Europe, of intellectual elites in classical Greece and Rome, and of the mercantile elites in the Mediterranean city states for much of the Middle Ages. Only in the past two hundred years have they become widely available and universally understood and used (Thrower, 1999).

From a political perspective, the attraction of maps is that while purporting to be neutral and comprehensive, they are actually inherently distorting and selective (Muehrcke, 1978). It is impossible to transfer a curved image onto a flat surface, or to include all the information available at any given point on the surface of the globe. Equally, all maps and plans have to exaggerate to some degree, otherwise significant features, such as rivers, roads, or buildings, would be so small as to be invisible and the main purpose of the map itself would be lost. As a result, cartographers have enormous licence to mould the images they create to reflect their own, or their paymasters', particular world view.

With small-scale maps, showing the whole world or individual continents, the choice of projection can fundamentally alter the image being mapped, an attribute that gave rise to Lord Salisbury's outburst cited at the beginning of this chapter. As can be seen from Figure 2.4, the Universal Transverse Mercator Grid hugely exaggerates the extent of the higher latitudes, while compressing the lower latitudes, thus sacrificing the true representation of area to the interests of relative location, because all compass bearings can be shown as straight lines. Equal area projections, on the other hand, accurately depict area, but distort where land masses sit in relation to one another, potentially giving rise to quite unjustified territorial associations. Even when cartographers are scrupulously careful to be specific about how a map is constructed, it is very difficult for the full consequences of any caveats to be taken on board by all the possible users, leading inevitably to misunderstanding and misconception.

An excellent illustration of the controversy that maps can generate is the argument over the value and accuracy of the world map devised by the German cartographer Arno Peters in 1974. He claimed that his projection illustrated equal area on a world map for the first time without undue fragmentation of the grid upon which it was drawn. What it certainly did was to visually emphasise equatorial and tropical latitudes at the expense of the higher latitudes. Whether the map was truly original or a modification of earlier projections was heatedly and rather inconclusively debated, but as far as political geography is concerned it focused on the intrinsic ideological content and power of maps (Crampton, 1994). Peters projection caught the public imagination and was used extensively to highlight the true extent of developing areas and their concentration in the tropics, thus underlining the scale of Third World poverty. Many non-governmental organisations concerned with overseas aid adopted the map to promote their cause, which was at the centre of a wider global political debate in the last quarter of the twentieth century about how to redress the economic disparities between the industrial and the non-industrialised worlds. The former are predominantly in the temperate latitudes, the latter predominantly in the tropics (Vujakovic, 1989). Whatever its cartographic merits and originality, Peters demonstrated conclusively to a wide audience that a map could be a powerful ideological weapon, capable of determining the terms of political debate in the most forceful manner.

Large-scale maps of smaller areas are less beset with problems of physical distortion, but the information included, or more often not included, may lead to equally significant misconceptions. The map can

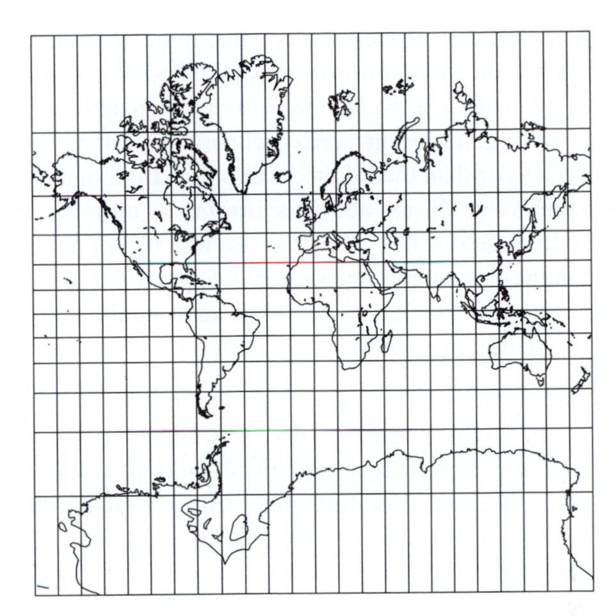

Mercator projection

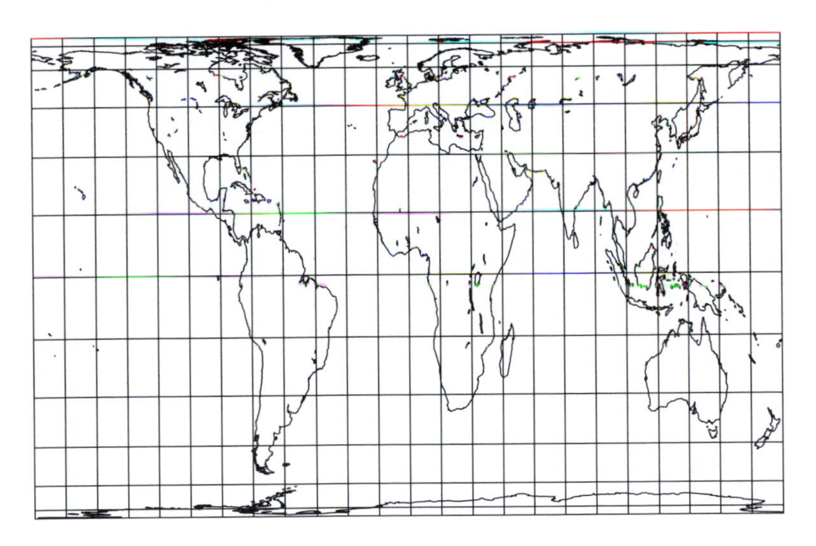

Gall–Peters orthographic projection

Figure 2.4 ***Different worlds: the globe projected using the Mercator and the Gall–Peters orthographic projections***

look completely different and assume quite different meanings, depending on what information is included or excluded, and in what form. Even the written language used on a map is politically significant. The spelling, for instance, may reflect the working language of the map-maker, rather than that of the local population; in the past this has led to centuries of confusion about the origin and meaning of place names, for example. Equally, it is often the case that two or more languages are current in a given area and selecting any one for use in a map elevates it above the rest, conferring a quasi-official status to an essentially arbitrary choice (Ormeling, 1983).

A good example of such arbitrary distortion on large-scale maps is the results of English-speaking surveys in Ireland in the nineteenth century. The surveyors used anglicised transliterations of Gaelic words to identify places and features, resulting in a plethora of names that had nothing to do with the true meaning of the unwritten original and its associations with place. The full extent of the distortion only became apparent when the maps were reinterpreted back into Gaelic, but by this time the language itself had all but died, so that the original meanings were in any case of little more than academic interest, as a new geography had effectively been created based on linguistic gobbledegook (Robinson, 1986).

It is both a strength and a weakness of maps that they exude an air of veracity. The information they contain tends to be taken at face value and widely believed, even though by their very nature maps must distort. One of the reasons for this is the unequivocal way in which cartographic messages are conveyed. Information is either present or absent and equivocation and uncertainty are hard to accommodate. Transition zones of all kinds are difficult to incorporate clearly into a cartographic format and imprecision, including political imprecision, tends to be masked. Disputed and undemarcated political boundaries provide excellent examples of the problems that can arise. Sinnhuber has shown how different national atlases in the former East and West Germany and in Poland depicted the boundaries between what were then the three countries in significantly different ways, each reflecting their particular geopolitical aspirations (Sinnhuber, 1964). Equally, there are still many national boundaries that have not been properly surveyed and continue to be disputed, notably in north Africa between the states spanning the Sahara desert, and in South America between the states along the spine of the Andean cordillera. However, few maps give any clue to the uncertainty and provisional nature of what is shown.

Maps, the state, and empire

Although governments and those with political power had always used maps as a means of demonstrating and confirming the control they sought to exercise over space, it was not until the beginning of the nineteenth century that map-making became fully absorbed into the apparatus of the state. A succession of wars seeking to establish unequivocally the extent of state hegemony in Europe, culminating in the extensive campaigns of the Napoleonic Wars between 1793 and 1815, which at one point seemed to promise a French state extending from the Atlantic into Russia as far east as Moscow, made it increasingly obvious that accurate and comprehensive information about land and territory would be a huge military asset. At the same time, governments were seeking to identify people with the state in a way that had never been attempted before and the emergence of the concept of the nation state encouraged military map-making and state-inspired political image-making to come together in a powerful alliance (see Chapter 3).

Nearly all states now have some form of national mapping agency, charged with producing accurate and up-to-date maps. The scale and sophistication of these documents varies enormously, but many still retain explicit reference to their military origins. The earliest and still one of the most comprehensive is the Ordnance Survey of the United Kingdom, which was founded officially in 1841 by the Ordnance Survey Act, but whose origins actually go back to 1791 (Skelton, 1962). Initially the goal was to survey the whole of the country at a scale of 1 inch to 1 mile (1:63,360), but the huge variety of uses to which the new maps could be put, ranging from civil engineering to town planning, as well as providing basic information for military purposes, quickly convinced the British government that something more detailed and at a variety of scales was required. In 1858 it was decided that across the whole country maps should be produced at a scale of 1:2,500 for cultivated areas, 1:10,560 for uncultivated areas, and at 1:500 for towns with a population of over 4,000. All other maps, including the 1 inch to 1 mile series, were to be derived from these basic surveys (Harley, 1975).

In the ensuing century and a half this huge project has suffered considerable vicissitudes as a result of changing political priorities and the need to incorporate technical advances in cartography. However, the whole of the United Kingdom has been surveyed and regularly updated at scales of 1:1,250 for major urban areas, 1:2,500 for minor towns and cultivated areas, and 1:10,000 for mountain and moorland. The maps provide a datum against which the nature of the national environment is routinely measured, but in a real sense they are no more accurate than their technically cruder predecessors. Users, as Harley has so forcibly pointed out, when considering the accuracy of maps must always first ask 'Accuracy for what?' (Harley, 1975, p. 159). Geometrical distortion and selectivity as to what information is included or excluded remain as fundamental to the cartographer's art as ever. As far as the Ordnance Survey maps are concerned one can look in vain for many military installations, excluded from the maps on the grounds of national security. A case in point are the absent military airfields in Oxfordshire and East Anglia, where the maps show agricultural land with no settlement at all. A simple piece of detective work, comparing the vacant areas on the map with the official population census statistics, easily uncovered the evasion, but it is a salutary warning that even the most apparently authoritative maps must be approached with a healthy dose of scepticism (Beckinsale, 1972).

Beyond the mapping of the national territory, cartography has long been used by states to legitimise imperial conquest. Surveyors accompanied the military as they ranged ever more widely across the globe seeking to acquire new lands. The maps produced were hugely influential in defining the extent of political hegemony, even though they included little or nothing about the detail of the lands they were depicting. They were geometric abstractions and Harley claims that 'The rediscovery of the Ptolomaic system of co-ordinate geometry in the fifteenth century was a critical cartographic event privileging a "Euclidean syntax" which structured European territorial control' (Harley, 1988, p. 282). From the rectangular land surveys in the United States, to the hurried cartographic annexations that accompanied the colonial carve-up of Africa, the imposition of an abstract geometry enabled land to be commodified without reference to past or existing use, or anything other than a two-dimensional geography. This depiction of a flat and featureless environment had the effect of absolving the new colonial governments of any sense of responsibility for the historical legacy of the lands they were annexing and, as a direct result, perpetrated some of the worst political injustices and atrocities in the history of civilisation. Native peoples, particularly in North America, Africa, and Australasia, were literally sacrificed to the colonial cause; their plight could be conveniently ignored, because the largely blank maps that colonial governments used as their main sources of reference meant that the native peoples were little more than an irritating impediment on the road to a manifest European destiny (Livingstone, 1994).

Maps and propaganda

Implicit in much of what has been said so far in this chapter is that maps have a great power to mislead, empowering those who make them by very effectively creating a preconceived desired image in the minds of map users. It follows that maps may also be used malignly as weapons of propaganda, using cartographic sleight of hand deliberately to mislead. There are numerous ways of doing this and the whole subject has been comprehensively and entertainingly reviewed by Mark Monmonier in his book *How to Lie with Maps* (1991). For present purposes, therefore, discussion will be confined to political propaganda, where the aim has been to create a deliberately misleading impression, with the express intention of influencing the terms of debate and justifying a particular course of political action.

One of the most blatant examples of such propaganda was the systematic and persistent overestimation of the scale of the Communist threat to Western Europe and North America. The deception was perpetrated in two ways. First, the fact that the Soviet Union and the People's Republic of China, the two major Communist countries in terms of both area and population, shared a common border stretching some 2,000 km and that most of the Communist satellite states in Eastern Europe also adjoined the Soviet Europe, enabled the Communist threat to be represented as a single monolithic block, threatening to overwhelm its smaller and fragmented western counterpart (Figure 2.5). Rarely, if ever, was any reference made to the vast areas of empty territory, punctuated by small constellations of settlement and development, that are the reality in this apparently undifferentiated region. Second, the fact that the greater part of the Soviet Union is located in the higher northern latitudes meant that, by using the Mercator projection to depict it, the impression could easily be given that its areal extent was much greater than was in fact the case. While it is perhaps tempting to claim that the prevalence of this louring image was purely accidental, it is interesting that since the break-up of the Soviet Union in 1989 a much more fragmented and diversified cartographic image of the territory of its former territory has begun to be displayed, emphasising, for example, the emerging regional power block in and around the Black and Caspian seas and the Caucasus Mountains that separate them.

A more frequently quoted example of maps and cartography being employed in a battle for hearts and minds is the extensive use of maps by the National Socialist government of Germany's Third Reich between

Figure 2.5 ***The Communist threat to the West depicted using the Mercator projection***

1933 and 1945. The driving obsession of this regime was to reassert Germany as a key European and global power after the humiliation the country had suffered after the comprehensive defeat of the Second Reich in the First World War. To this end, it used maps in the most tendentious fashion to illustrate how a reduced and beleaguered Germany was encircled by hostile states, and how territorial expansion was a necessity if it were to escape from the stranglehold. The images were published widely, not only in Germany, but also in the United States in a concerted campaign to enlist support for Germany's cause. Not surprisingly, much of the steam ran out once America joined the Second World War on the opposite Allied side in 1942 after Germany's ally Japan attacked the American fleet in the Pacific Ocean at Pearl Harbor.

Finally, it would be quite wrong to give the impression that maps and cartography have only been used for political ends in national and international conflicts. Some of the most successful cartographic forays have been in support of local issues, where the hearts and minds of the electorate need to be won. In the 1970s, there was widespread concern in England and Wales that the national parks were about to be overrun as part of a mass invasion of the countryside from the towns and cities. As part of its campaign to avert this the Dartmoor National Park published a map, showing the numbers of people who would be within two and a half hours' driving time of its boundary once the M5 motorway was complete. The clear message was that millions would be using Dartmoor as their main recreation destination. Crucially, the map made no mention of all the alternative destinations, including other national parks, for tourists from Birmingham, Bristol, and south Wales, many of them much less than two and a half hours' driving time away. Despite this fundamental omission, the map became a powerful weapon in the argument for more restrictive tourist policies for Dartmoor, attracting widespread local political support.

The future of maps – mapable information on demand

Traditionally, maps and access to map information have been tightly controlled and it is this that has been the source of their enduring political influence. Even with the advent of national cartographic surveys, such as the Ordnance Survey in the United Kingdom, producing maps at a variety of scales on open sale to the public, a degree of control has always been exercised over what is and is not shown. In the past two decades, however, technological advances mean that the sources of that control have all but disappeared and virtually all visible information about the surface of the globe is potentially available on demand. The agent of this fundamental change is a combination of developments in geographic information systems (GIS) and satellite imagery, which means that any part of the world can now be photographed with a sufficient degree of resolution for detailed maps to be constructed and with little or no restriction on who has access to the data.

It is impossible as yet to say what the long-term consequences of this revolution will be, but it is certain that cartography will no longer be the exclusive preserve of powerful elites, nor will there be the same restriction on what is mapped, since an immeasurably wider range of information will be available to put into map form (Cho, 1998). Nevertheless, it is unlikely that the democratisation of map-making and cartography that information technology and the satellite age have ushered in will result in maps losing their political clout and effectiveness. Indeed, it seems likely that their power will increase as more people are able to deploy them in support of their own political agendas.

Key themes and further reading

The main theme of this chapter has been the representation of space. Readers should now have a clear idea of the meaning and political implications of territoriality – identifying, laying claim to, and defending specific areas of land. They should also understand the process of map-making and the importance of, and distinction between: symbols for demarcation, systems for codifying claims, and agency for enforcement. The role of maps and cartography are central to the whole process, but the limitations and partial nature of maps as sources of information should now be clearly apparent. Historically, maps have been a weapon which powerful minorities have used to consolidate their control over space. The way in which maps can be used for propaganda purposes to distort the real world should now be understood. The impact of GIS and other computer technologies on the availability and use of spatial data is currently making maps of all kinds much more widely available, but it should be appreciated that this does not necessarily remove their capacity to distort.

The whole concept of territoriality has been explored in detail by Robert Sack (1986) in his book, *Human Territoriality: its theory and history*. It is jargon-free and readable, without becoming mired in the more difficult aspects of a complex concept. To find out more about maps as symbols of power and patronage, the many books and journal articles by J. B. Harley are essential reading. One of the best, and an excellent introduction to the political geography of maps, is 'Maps, knowledge, and power' (1988) which is published in *The Iconography of Landscape* edited by D. Cosgrove and S. Daniels. The role of maps and spatial images of all kinds is changing rapidly as a result of satellite imagery and recent developments in IT. These issues are comprehensively discussed by G. Cho (1998) in his wide-ranging book, *Geographic Information Systems and the Law*.

3 The idea of the state

> Civil power, properly considered as such, is made up of the aggregate of that class of the natural rights of man, which become defective in the individual in point of power, and answers not his purpose; but when collected to a focus, becomes competent to the purpose of every one.
>
> (Thomas Paine, *The Rights of Man*, 1791)

The concept of the state

States and the idea of the state have traditionally formed the cornerstone of political geography, providing the key terms of reference for explaining the distribution and exercise of political power. Even though their continuing relevance has been increasingly questioned in recent years as regional and global institutions have begun to evolve which are less fettered by the constraints of state control, states remain central to understanding the social and economic dynamics of the political landscape. As Thomas Paine so cogently argued more than two hundred years ago, natural rights cannot be guaranteed by individuals on their own; they require the support of a collective, civil, authority if they are to be a reality. He was writing at the time when the whole idea of state identity being invested exclusively in the person of a monarch, or some other absolute ruler, was crumbling in the face of the rise of capitalism, the spreading urbanisation of the population, and popular demand for the people to have a greater direct say in government.

The role of the state, or more precisely its more confined antecedent the city state, was first articulated by the philosopher Aristotle in ancient Greece in the third century BC (Nicholson, 1984), but it is only since the advent of the new economic order based on capitalism and industrialisation at the end of the eighteenth century that states have become established in anything approaching their present form (Box 3.1). In pre-agrarian societies, few of which still function as such in the modern world, tribal loyalties were the main force for social cohesion. Groups of nomadic hunters and gatherers were too small, scattered, and ephemeral to have the time or the need to develop formal political institutions, though complex, sophisticated, and rigidly enforced codes of behaviour invariably underpinned the stability of daily life (Cohen, 1978).

Box 3.1

Aristotle's city state

The Greek philosopher Aristotle developed his theory of the city state in about 350 BC in his book, *Politics*. Based on his experience of life and government in Athens, he argued that a natural logic dictated that societies should have government, or political rule, and that government, in its turn, led inevitably to the emergence of the city state. He demonstrated, in terms that would be highly contested today, how city states gradually grew out of much simpler communities. First, individual humans combined in pairs because they could not survive alone: men and women came together to reproduce; the master and slave stayed together for mutual self-preservation. The master used his intellect to rule, while the uneducated slave used his physical strength to labour. Second, the household unit arose naturally out of these primitive communities to serve domestic and economic needs. Third, groups of households quickly combined to serve higher order needs, resulting in villages. Finally, villages inevitably merged to form city states, complete and self-sufficient communities, which originated as the logical culmination of a natural order, but which survived because they are the best guarantee of a good life for citizens.

The settled agrarian societies that followed, by contrast, rapidly developed a need for institutional coherence and organisation to sustain their more elaborate social order. Coherent and defined territories became essential and the production and marketing of agricultural, and other, surpluses were a measure of their success. This in turn presupposed the emergence of literate and educated elites to manage the more complicated social and economic relations, both within the society and with the wider world beyond, which led inexorably to a greater division of labour and to hierarchical social structures. Access to territory and the ability to exercise exclusive control over it also became increasingly important issues, leading inevitably to conflict, attempts at conquest and, in extreme situations, war.

Nonetheless, most agrarian societies were largely self-sufficient and self-contained, though frequently embedded within a loose overarching polity, such as the Roman Catholic Church of the Holy Roman Empire in early modern Europe. Gellner (1983) has emphasised that political units in the agrarian age varied enormously in both size and kind, but suggests that they may be divided broadly into two types. On the one hand, there were city states and rural, peasant, communities largely running their own affairs with high levels of political participation, at least for men. On the other, there were extremely large territories or empires, controlled by a single dominant force, with power concentrated at a single point. Frequently, of course, the two coexisted side by side, with a dominant, but remote, central authority operating alongside largely autonomous local units. Indeed, it can be argued that somewhat similar arrangements still predominate in many parts of the world, notably in countries such as Russia and the People's Republic of China, where the central political authority of the state is thousands of miles removed from the everyday lives of many people and communities. They organise their affairs in a largely self-sufficient manner, in some cases virtually dispensing with money as a medium of exchange, relying instead on extended forms of barter, which insulate them still further from the wider world beyond the confines of their own neo-feudal social and economic systems.

The modern state is a product of the post-feudal order, in which the state gradually took over increasing responsibility for managing the process of socialisation. The origins of this fundamental change in Europe date back to the Reformation in the mid-fifteenth century, when the monolithic ecclesiastical hegemony of the Roman Catholic Church began to break down, to be replaced by more localised and independent Protestant alternatives. In England the rupture was also closely identified with the state, in that the monarch, Henry VIII, was the driving force behind the break with Rome and the establishment of the overtly national Church of England (Box 3.2). It meant that for the first time in the modern period church and state were explicitly brought together in a unified entity, with the monarch, the head of state, also leader of the official national church.

It was, however, the establishment of capitalism as the dominant form of economic organisation three hundred years later, alongside industrialisation, which really heralded the emergence of the intricate

Box 3.2

The English Reformation

The Protestant Reformation only gathered ground slowly in England in comparison with some other European countries, such as Germany and Switzerland. Somewhat ironically, it did eventually begin to gain ground during the rule of King Henry VIII, a very strong defender of Roman Catholicism during the early years of his reign, when the Pope actually bestowed on him the coveted title, Defender of the Faith.

It was his increasingly desperate determination to be rid his first wife, Catherine of Aragon, that drove Henry VIII to decide to split away from Rome, which refused to countenance divorce. In 1532, legislation was passed through Parliament limiting the influence of the papacy in England and making the monarch the Supreme Head of the Church. Once having successfully effected his divorce, Henry VIII went much further and took control of the majority of the property of the Church through the dissolution of the monasteries.

There was little popular enthusiasm for the change in religious allegiance, but for the most part people acquiesced, in some cases encouraged by the redistribution of Church property in the wake of the confiscation of monastic lands. However, after the death of Henry VIII in 1547, more active steps were taken to consolidate the position of the new Protestant Church. The regency government representing his under-age successor, Edward VI, set in train a determined programme of reform, resulting in a new Prayer Book and a new order of service, as well as the removal of most of the physical artefacts of Catholicism from the England's churches.

After only six years on the throne, Edward VI died and was succeeded by Mary I, who, in her reign between 1553 and 1558, effectively reversed the whole Reformation, returning England to Catholicism. Her successor, Elizabeth I, then determinedly set about re-establishing Protestantism and gradually, during her long reign, the new religion took an increasingly firm hold. When she died, childless, in 1603, the dominant position of Protestantism was further secured when the throne passed to King James VI of Scotland, who became James I of England as well, creating for the first time a joint Protestant kingdom across the two countries.

patchwork of independent states that now characterise the world political map (Hobsbawm, 1990). Urban–industrial society encouraged an ever more sophisticated division of labour and a growing dependence on technologically advanced communications systems to link the varied industrial skills into a coherent and viable national economic whole

(Harvey, 1982 and 1985). The actual production of goods and services became increasingly separate from the means of production by the commodification of the workforce. Labour became just another tradable element in the marketplace, rather than being ineluctably tied into a rigid feudal system. In theory, workers were able to sell their skills wherever they liked, to the highest bidder, even though in practice, of course, all kinds of constraints, such as tied company housing, ensured that the market in labour was far from free for many. Nevertheless, there was a new flexibility that would have been unthinkable within the more rigid framework of pre-industrial agrarian society. Also, the competition between employers at all levels helped prevent, though not totally eliminate, the creation of exclusive monopolies. Although many major industrial cities in North America and Europe were dominated for many years by a single employer, such as Pittsbugh and the US Steel Corporation, Essen in Germany and Krupp A.G., which effectively nullified the advantages of a commodified labour market, over time these monopolies loosened their grip, as the cities grew and the employment market diversified through the influx of new firms and companies, all competing for labour.

An important feature of the urban–industrial, capitalist environment was the fiction it created that the economic and political environments were separate, the market supposedly ensuring that economic interests were fully engaged competing with each other, leaving states to provide the political shell within which the economy functioned. It was a fiction because the supposed separation was demonstrably false. The economic leaders always sought to manipulate the political system to further their own interests and, to this end, spent much of their time and energy trying to establish themselves in positions of political power and influence. Equally, the state had a quite legitimate interest in ensuring that its economic infrastructure flourished, not least in a highly competitive international environment. Indeed, one of the most serious criticisms levelled against the capitalist system has been that, after a period of sustained success in expanding world trade throughout the greater part of the nineteenth century, it failed to control competition between states and establish a stable international order capable of preventing two world wars (Carr, 1968). At the beginning of the twenty-first century, it is now under attack from a different direction, for failing to prevent the development of a global trade system that is manifestly unfair, consigning the bulk of the global population to a life of poverty (see Chapter 11).

Recognising the need to devise new forms of political infrastructure, which incorporated a much greater level of explicit general consent, states in the modern era began to provide for popular representation in government and to formulate and adopt written constitutions on behalf of all their population. The USA is often cited as a model in this regard, with its 1789 Constitution, proclaiming freedom and equality for all, although it was actually signed by just 39 male delegates from 13 of the then states along the eastern seaboard of North America. Such lofty ideals were widely praised and copied, but actual participation in government was still invariably strictly controlled, based on age, gender, wealth, or a combination of these, effectively excluding all but a privileged minority. The constitutions themselves defined, with varying degrees of precision, the nature of the state, specifying, for example, the official languages and in some cases the national religion. They also set out the mutual responsibilities of the state and the citizen and how they were to be managed, paving the way in the process for a whole raft of state-run services. Communications, such as the postal service, became a matter for the state, as did the provision of social services like public health and education, and the maintenance of public order through the creation of an effective national legal system and police service.

An important by-product of the dominance of the capitalist economic order and its attendant centralisation and state control has been a growing sense of nationalism. It is a binding political force whose initial roots can be traced back clearly at least as far as the concept of a chosen people in the Old Testament (Hastings, 1997; Davie, 2000). In the modern world, however, nationalism has achieved unprecedented importance and is characterised by a close identification of the population, and popular ideals, with the state and the values of the national government, and by the spread of democracy as the means by which these are communicated to government (see 'The state apparatus' in Chapter 4). People become citizens, active participants in the process of nation-building and begin to define themselves in terms of the state, becoming in the process American, British, French, or whatever, ultimately with passports and identity cards to prove their right of citizenship. This in turn gives rise to the concept of the nation state, whereby the organs of the state penetrate and control the civil society, to a point where they are one and the same (Mann, 1984; Giddens, 1985). Nationalism became an increasingly potent force throughout the nineteenth century and was viewed initially as a positive force for its fostering of an integrated political order, but rapidly came to be seen less kindly as a focus for interstate rivalries and for the territorial ambitions of European states in particular. The horrific conflicts of the late nineteenth and twentieth centuries, global in scale in the cases of the First and Second World Wars, are now attributed in no small part to the malign influence of nationalism and the need for states to be able to demonstrate to each other their superior power and influence (O'Loughlin and van der Wusten, 1993).

It is somewhat ironic that just at the time when states were striving most energetically to show off their virility and effectiveness, the limitations of the state idea were also becoming all too painfully apparent. In fact, the concept of the state as a self-contained entity, bringing the whole of economic, political, and social life into a single coherent whole, was always an illusion. In a critique of the nature of power in the nineteenth century, Carr argues that 'it was precisely because economic authority was silently wielded by a single highly centralised autocracy that political authority could safely be parcelled out in national units, large and small, increasingly subject to democratic control' (1945, p. 87). The implications of a lack of effective democratic control over the executive were well understood by political analysts at the time, including Karl Marx (Wheen, 1999), and have been repeatedly demonstrated in the subsequent years, threatening to undermine some of the most ambitious political integration projects, such as the European Union.

The so-called globalisation of the world economy and the limited ability of states to determine their economic and political destiny is, therefore, not a new phenomenon. It is the latest manifestation of a tension that has always existed, but which has tended to be downplayed by governments for reasons of national pride. In the political arena, international institutions to counter the darker side of nationalism, under the umbrella of the League of Nations after the First World War and the United Nations after the Second World War, have been extensively developed, though with limited success. In the economic arena, free-market economic entities largely bypassing state control, such as supranational companies and global financial institutions, have been facts of life for most of the twentieth century (see Chapters 11 and 12).

The nature of the challenge to state authority has now moved on to a different plane in the twenty-first century. The so-called information age has gone a long way towards liberating the educated and the wealthy from many of the constraints of state control, with widespread access to information technologies and personal communications networks that transcend and ignore national frontiers (Castells, 1997). Global links of all kinds can now be made by just a few clicks on a computer mouse and, with convertible currencies increasingly the norm, access to worldwide markets for goods and services is available on an unrestricted individual basis in a matter of minutes, or even seconds.

Nevertheless, it is important to stress the limits on what is available. For the majority of the world population, languishing in poverty and struggling to find shelter and enough to eat, the benefits of individual access to information technology are an incomprehensible joke, with little or no relevance to their daily lives. Having said that, it is also true that in relative terms information technology is cheap technology, so that it is rapidly permeating societies largely bypassed by earlier technological revolutions. In the countries of south-east Asia, for example, the high levels of IT literacy are doing much to fuel the rapid transformation from a semi-feudal to an industrial, capitalist society within the space of a generation. However, it is also the case that although state control may have been weakened, it has by no means been eliminated. It is still possible to manage access to the information highway, either by charging for its use or imposing restrictions on access. As a result, there are big variations between even the most developed industrial societies in the extent to which the internet is used. In the USA, where there is unmetered access via the telephone system, internet usage is virtually universal, whereas in Europe access is generally metered and levels of usage substantially lower.

In practice, there have always been significant limits on the ability of states to manage the full range of matters that concern their peoples. War in the twentieth century may have raised the alert, but other quite different issues, such as the problems of global environmental degradation and the availability and distribution of food and other key resources, have subsequently underlined the message (Deutsch, 1981). States remain an essential and ubiquitous element in the world political order, but they are only part of a complex hierarchical structure within which power relations are in a constant state of flux.

The spread of states

At the beginning of the third millennium the world political map comprises some 180 independent states, varying hugely in both area and population. The most extensive is the Russian Federation, covering 17,075,400 sq km; the most populous the People's Republic of China, with an estimated 1,400 million inhabitants; and the smallest, in terms of both area and population, the state of the Vatican City, a micro-state located entirely within the Italian city of Rome, covering 0.44 sq km and with only just over 1,000 inhabitants. This complex network embraces the land masses of all the continents with the exception of Antarctica, as well as incorporating increasingly large tracts of the oceans; yet it is almost entirely a product of the past two centuries.

The age of mercantilism

In 1800, nation states barely existed in the form they are recognised today and the few that did were almost exclusively concentrated in Europe. Nominally, the bulk of the globe was encompassed politically by mercantilist empires, with the land held in the name of some far-off monarch, or some other absolute feudal ruler. In reality though, these were often little more than a European conceit, consisting of scattered coastal trading posts, with only intermittent and nominal links to their European sponsors. The largest of these, both territorially and in terms of the scale of trade, was the British Empire (Figure 3.1), ranging over much of North America, the Indian subcontinent, and Australia, but other European states, including Denmark, France, Portugal, and Spain, also had substantial overseas territories over which they claimed sovereignty, with all its associated exclusive rights. The reality everywhere, however, was that imperial control was for the most part nominal and the indigenous peoples were largely unaffected and able to continue their lives much as they did before the coming of the European enlightenment.

The imperial model was not exclusively European. The Ottoman Empire, with its heart in the west of modern-day Turkey, covered large parts of north Africa and Asia Minor (now more usually incorporated into the wider region of the Middle East), as well as extending deep into the Balkan peninsula in south-east Europe. The Russian Empire covered most of northern Asia and extended across the Bering Sea into North America in the east, as well as encompassing most of the Caucasus in the west, thus giving Russia too a toehold in Europe. The Chinese Empire already formed a monolithic bloc, covering the greater part of the south-east Asian mainland, much as it still does today. Finally, the USA was beginning its dramatic westward expansion across the central part of North America.

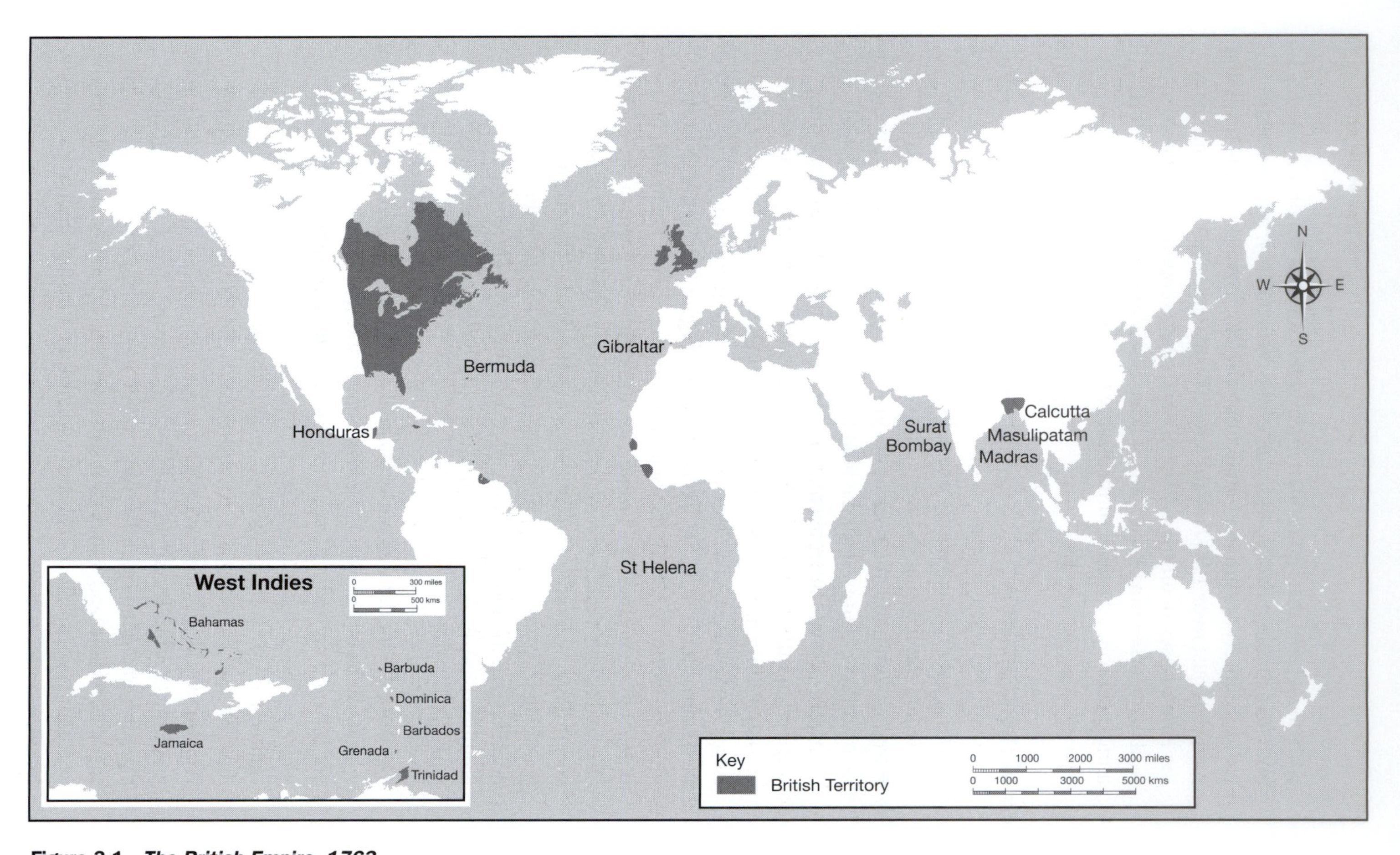

Figure 3.1 ***The British Empire, 1763***

Elsewhere the political order was much more self-contained and isolated, even though other significant political entities were well established. In east Asia, Korea and Japan both maintained sophisticated and flourishing cultures; in central Asia loose-knit and fluid societies, including the Mongols, continued to dominate in 1800 as they had for centuries previously; Persia was a pivotal presence between Europe and Asia; and in Africa, Abyssinia and Morocco were the best-known elements of a wide range of what can broadly be called monarchies spread throughout the continent. A unifying thread linking all these disparate and widely separated entities was that they were beyond the direct scope of European, or any other, external imperial hegemony.

The tide of nationalism

The latter part of the eighteenth century witnessed a fundamental change in popular attitudes to government. Across the world there was a growing restlessness and resentment against feudal absolutism, especially when it was exercised from a faraway continent in little more than name. In North America, a quarter of a century of struggle to oust British rule culminated in the establishment of the USA in 1783. In Europe, the huge upheaval of the French Revolution took proper root in 1789, presaging more than two decades of war, bloodshed, and change, which completely recast the political landscape of the continent.

Everywhere the overt goal was for more representative government, which would be responsive to the emergent tide of nationalism, as discussed at the beginning of this chapter (see also 'Nationalism and self-determination' in Chapter 5). Somewhat surprisingly, the temporal sequence for the founding of the newly independent states that emerged as a result of the massive outbreaks of revolutionary zeal did not quite mirror the fervour of the revolutions themselves, especially in Europe. It was in the Americas where a new order first became firmly established. Following its success in establishing itself as a republic, the USA was eager to see an end to European colonial rule throughout the whole of both North and South America. For their part, the European colonial powers, Spain, France, Portugal, and Britain, were unable to sustain simultaneous wars on either side of the Atlantic, especially as the internal political structure of their pre-revolutionary states was being completely recast. From the beginning of the nineteenth century, France, Portugal, and Spain all began a rapid withdrawal to their European heartlands, following in the wake of the British after the American Revolution.

After Mexico became a republic in 1823, republican fervour swept across the greater part of the Americas throughout the rest of the century. In some cases, the newly established republics were relatively short lived, being incorporated into larger neighbours after a relatively short interval. The State of Texas, for instance, fought a 10-year intermittent war to separate itself from Mexico, finally becoming independent in 1836. However, a growing stream of European immigrants from the north promoted ever closer ties with the USA and, in 1845, Texas became another state in the growing American Union. The republican movement proved irresistible across the Americas and by the turn of the twentieth century virtually the whole of Central and South America was governed by nineteen independent republican states.

There is no doubt that this avalanche of state-building was much encouraged by the distraction of the former European colonial powers and a determination on the part of the USA to allow them no opportunity to re-establish a colonial foothold in the Americas. Under the terms of the Monroe Doctrine (see Chapter 9), it committed itself to providing naval protection against any threat to the independence of the newly founded republics. Ironically, the only major area that has remained untouched by the republican tide is Canada, which shares a 5,000-km long border with the USA and still retains the British monarch as its nominal head of state, a final remnant of the traditional colonial era.

In Europe, nationalist fervour was translated into new republics rather more slowly. The Congress of Vienna was first convened in 1814 to re-establish political order after the Napoleonic Wars, and the participants – the four major victors, Britain, Russia, Prussia, and Austria, together with defeated France – were certainly not in any mood to adopt the populist ethos of the French Revolution. The decisions made by the Congress actually put the clock back, because of the reactionary way in which they tried to revive Europe's traditional attachment to monarchy, rather than embracing the new republicanism (Davies, 1996, p. 762) (see Chapter 9). Although the deliberations led to the outlines of the modern political structure of the Low Countries and Scandinavia, as well as Switzerland, they also resulted in the imposition of a string of new, or resurrected, monarchies. The most notable example was the Kingdom of the Netherlands (covering both the modern-day Netherlands and Belgium), but in the future unified states of Germany and Italy there also remained a plethora of kingdoms, duchies, principalities, and papal states whose government was far removed from the popular aspirations of the revolution.

Nevertheless, the revolutionary zeal was only temporarily quelled and soon began to reassert itself. By the middle of the nineteenth century, programmes of political reform, both with and without accompanying revolutions, were widespread across Europe. In Great Britain and Ireland the changes were internal and largely peaceful with the Reform Act (1832) initiating what was to become a fundamental redistribution of political power away from the monarch and a landed elite to the people as a whole.

Elsewhere, territorial redistribution and wholesale political change were required, though it mostly amounted to a severe curtailment in the powers of hereditary rulers, rather than in their wholesale replacement by republics. The patchwork of political units in both Germany and Italy was gradually merged, so that by 1870 it formed two somewhat precariously unified political entities, one led by an emperor, the other by a king. Spain survived the loss of the greater part of its empire in South America and the Pacific to remain an independent kingdom, even if one riven by regional discord. The Netherlands and Belgium, united under a single monarch in 1815, became two separate monarchies in 1830.

The most radical changes in Europe occurred in the Balkan peninsula, where the Turkish Ottoman Empire progressively disintegrated throughout the nineteenth and early twentieth centuries (Glenny, 1999). The volatile mix of different religions and languages ensured that the whole process of change, undoing more than three centuries of Turkish rule, was both violent and chaotic, creating an unstable mixture of mostly small monarchies, which struggled to produce any real sense of national unity. Greece, Romania, Bulgaria, Serbia, Montenegro, Bosnia Herzogovina, Albania, and Turkey itself all trace their modern roots back to this era and all have experienced substantial change in the process, many seeing their monarchies dismantled to make way for republics. With the exceptions of Greece, Turkey, and Albania, these in turn were enveloped by Soviet Communism during the Cold War, between 1947 and 1990, losing most of their effective political independence, only to re-emerge after varying degrees of further inter-ethnic bloodshed in 1989, after the collapse of the Soviet Union (see 'Managing difference' in Chapter 5).

The age of empire

Just as all these newly independent states were emerging in the Americas and Europe, the major European powers at the end of the nineteenth century were aggressively vying with each other to establish extensive overseas empires in Africa, Asia, and Oceania. Britain consolidated its control over the whole of the Indian subcontinent, much of Africa south of the Sahara, and Australia and New Zealand (Figure 3.2). At its zenith, before the outbreak of the Second World War in 1939, it was a common boast and, as it subsequently turned out a vain one, that the sun never set on the British Empire because it extended over so much of the globe. France controlled much of south-east Asia, including present-day Cambodia, Laos, and Vietnam, in what was then known as Indo-China, as well as large territories in north and north-west Africa. Germany had colonies in both south-west and east Africa, as well as smaller interests elsewhere. Belgium claimed most of the Congo basin in central and west Africa, while the Netherlands controlled most of the Indonesian archipelago. Portugal, despite leaving its largest colony, Brazil, in 1889, retained Angola and Mozambique, large colonies in south-west and south-east Africa respectively. Italy controlled Abyssinia and other extensive territories in north Africa. In all cases, these were just the major territories and they were supplemented by these imperial states claiming jurisdiction over scores of oceanic islands across the world.

Elsewhere, the USA consolidated its grip on much of North America by incorporating all the lands from the Atlantic to the Pacific oceans, south of the 49th parallel of latitude and north of the Mexican border (see 'Environmental determinism and the state' in Chapter 8). North of the 49th parallel, the British converted their relatively small and widely separated colonial holdings into present-day Canada, which stretches across the North American continent from coast to coast, as well as extending northwards nearly as far as the North Pole.

Superpower hegemony and the spread of the nation state

The fundamental redrawing of the world political map that occurred during the nineteenth century was made possible in no small part because powerful states were able to acquire vast global empires at the same time

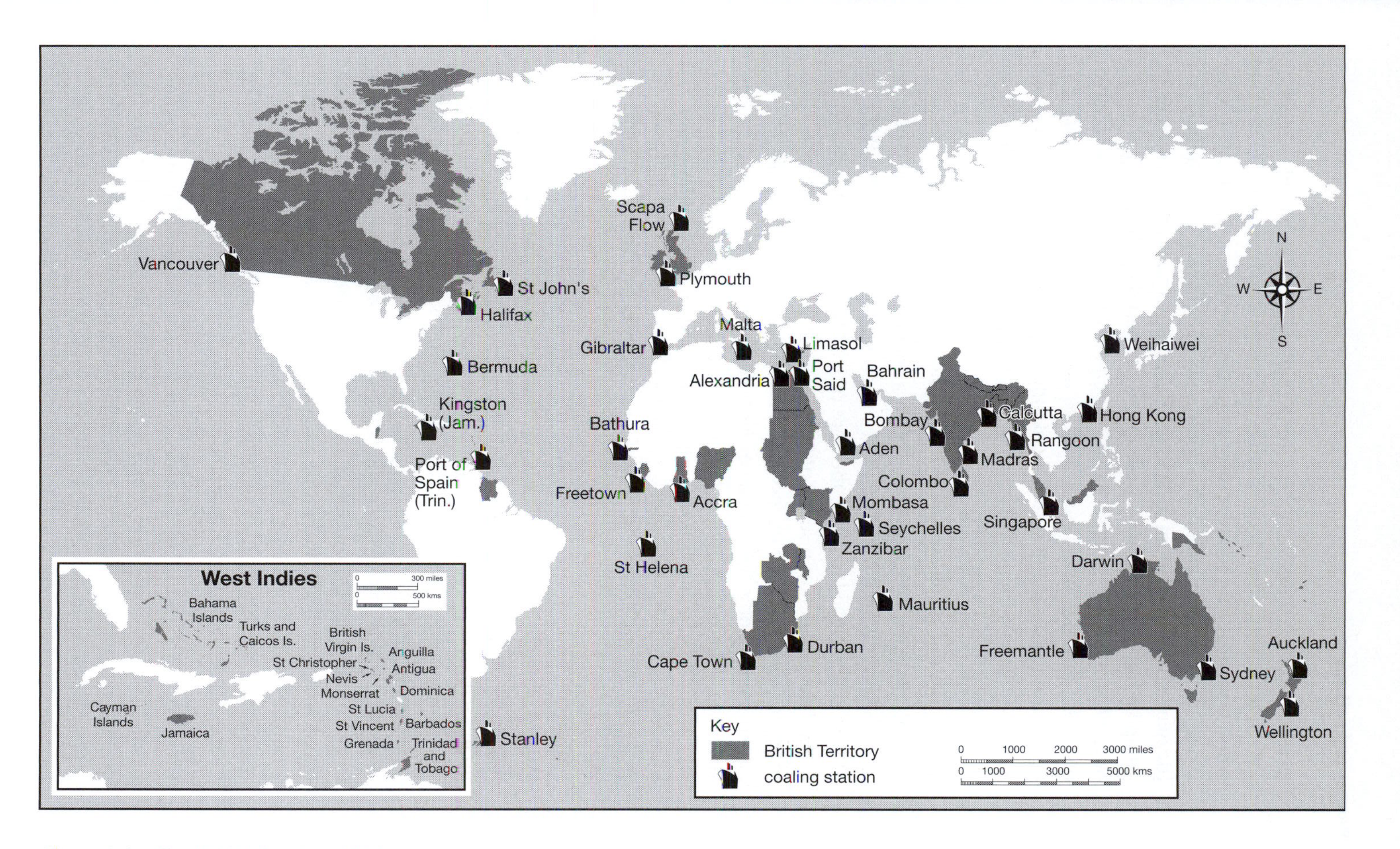

Figure 3.2 ***The British Empire, 1914***

as a host of new states were becoming politically independent (Cohen, 1964). It enabled them, by and large, to compete for extensive hinterlands without laying claim to the territory of other modern states (Carr, 1968). It was, however, a situation that could only ever have a very limited lifespan, since the amount of land available for expansion was always going be finite. Those limits were cruelly exposed by the First and Second World Wars, both of which were partly caused by irreconcilable national territorial ambitions in Europe and east Asia. The two wars also exposed the lack of forethought amongst the imperial powers about how a competitive global order might be managed and controlled through new international agencies.

The limitations first became fully apparent in the negotiations leading up to the treaties agreed to re-establish a stable global political order after the First World War. The most important of these was the Treaty of Versailles, signed in 1919, the core of which essentially revolved around the creation of a series of new states in Europe whose independence would be underwritten and protected by a global governmental body, the League of Nations. The boundaries of the new states were drawn largely on the basis of supposed national coherence, which in effect amounted to a rather crude assessment of ethnic and linguistic unity. The new states formed a swathe, stretching from the Baltic Sea in the north of Europe to the Black Sea in the south-east and the Aegean and the Mediterranean seas in the south-west. Modern-day Finland, Estonia, Latvia, Lithuania, the Czech and Slovak republics, Poland, Austria, Hungary, Romania, Bulgaria, and Albania all essentially date from this period. However, although the League of Nations was set up in 1920, it proved to be almost completely ineffective as a source of worldwide political authority. The United States refused to join and the emergent Soviet Union was not invited, so that it was left to a militarily weakened group of European states which were members of the League to try to safeguard the territorial integrity of a host of fledgling national democracies both within Europe and beyond. It manifestly failed and, as a result, rapidly became discredited in the face of malign imperial marauding, notably by Italy in Abysinnia and Germany in both the then Czechoslovakia and the Rhineland.

Elsewhere in the world, the former imperial territories of the two main losers from the First World War, Germany and the Ottoman Empire (the precursor of modern Turkey), were reapportioned amongst the victors, as protectorates or other similar sanitised formulations for the reality of colonialism. In the Middle East, Iraq, Syria, the Lebanon, and Jordan emerged in their present form from this process, as did Tanzania, Namibia, and Cameroon in Africa.

The territorial mix was, therefore, inherently unstable. More than anything else, it was the failure to provide guarantees of security for the nation states created after the First World War that led directly to the Second World War, geographically a much more truly global conflict. The aftermath of that second war, involving directly all five of the populated continents, not only extended dramatically the number of nation states at the expense of the traditional European empires, it also redefined the whole concept of empire in a less direct form of superpower hegemony. The new system has been legitimised and increasingly, though very patchily, underwritten by the United Nations, the global governmental organisation that succeeded the moribund and largely defunct League of Nations in 1945. All independent states may apply to become members and most have done so, seeing membership of the United Nations as a hallmark of their own legitimacy (see Chapter 12).

After the end of the Second World War in Europe in 1945, the fundamentals of the nation state system created by the Treaty of Versailles remained intact, albeit with the loss of a number of states to the Soviet Union in the east (Estonia, Latvia, and Lithuania) and the substantial redrawing of political boundaries throughout Eastern and Central Europe. Imposed upon this map, however, was the fault line of the Iron Curtain and the louring confrontation of the Cold War between the United States and its allies in the west and the Soviet Union in the east (Box 3.3). Elsewhere, in Africa, south-east Asia, and India, decolonisation progressed apace throughout the 1940s, 1950s, and 1960s as the European colonial powers found it increasingly difficult to sustain their protective role in the face of internal opposition from ever more confident nationalist movements and external competition for influence from the USA and the Soviet Union, the two Cold War superpowers, with intermittent additional incursions from a third, China.

The other continents were less directly affected. Australasia and North America did become more independent of their deeply rooted European connections and saw, and continue to see, a growing resentment of the colonial ethos amongst their populations, but for the most part this has been expressed through a more evolutionary process of internal change, reflected in a greater general acceptance of the claims by native Americans, Aborigines in Australia, and Maoris in New Zealand that

Box 3.3

The Iron Curtain

The Iron Curtain was the dividing line between Soviet Communist-controlled Central and Eastern Europe and Western democratic Europe for more than four decades in the mid-twentieth century. The term was first coined by the British wartime Prime Minister, Sir Winston Churchill, in a speech on 5 March 1946 at Fulton, Missouri in the USA. He defined the Iron Curtain as a line running from Stettin on the Baltic to Trieste on the Adriatic. He pointed out that a host of European capital cities were now firmly within the Soviet sphere of influence: Warsaw, Berlin, Prague, Vienna, Budapest, Belgrade, Bucharest, and Sofia. The clear implication was that Europe was severely diminished as a result. Subsequently, the definition of the Iron Curtain was modified to represent the whole of the dividing line between the Communist and free worlds in Europe, from the Northern Cape on the Norwegian–Finnish border in the north to the Greek and Turkish frontiers with Communist-controlled Yugoslavia and Bulgaria in the south.

The Iron Curtain has been described as 'one of the most powerful geographical barriers in continental Europe' (Blacksell, 1981, p. 15) and it provided a most durable political dividing line until the implosion of the Soviet Union in 1989/90. Since then, the Iron Curtain has become a historical curiosity and increasingly irrelevant to the geography of Europe. All vestiges of its former significance disappeared with the expansion of the EU to include eight former Communist states in 2004.

their rights be recognised. In South America too the process has been essentially one of internal change, though often accompanied by bitter and violent political conflict, fomented by indirect superpower involvement.

The collapse of Soviet Communism and in particular its economic system in the 1980s has had a profound impact on the state system (Fukuyama, 1992). Many states were cast adrift from their traditional Soviet superpower embrace, but were not immediately absorbed by the US alternative. As a result, a far less predictable political mix has evolved, with independent nation states very much the preferred alternative, but with much more volatile and fickle political allegiances. It makes for an unstable political environment, but one where nationalist sentiments are increasingly significant. Small states, such as the group of independent states that have painfully emerged from the wreckage of the former Yugoslavia since 1990, often feel able to challenge large and much more powerful ones directly, and frequently they use the threat of political instability and the danger of it spreading as a bargaining counter. Equally, the uncertainty and instability often encourages discontented minority populations within existing established states to press their claims for independence even more vigorously. In short, nation states look set to proliferate in the more deregulated world of the early twenty-first century.

A typology of states

In the discussion so far little attempt has been made to distinguish between states, with the implicit implication that they are more or less uniform. Obviously at one level this is patently untrue, since states vary in terms of their size, population, location, and many other physical characteristics, but it is also the case that they vary significantly in terms of internal organisation as well and that there are well-defined trends in the way states have developed over the past two hundred years.

Essentially there are three broad categories of state governance that have evolved: monarchies, colonial dependencies, and republics, each covering in detail a spectrum of different arrangements. Monarchies include kings, emperors, tsars, princes, shahs, and many more, but are all regimes where dynasties rule as of right. Colonial dependencies cover all those territories where the responsibilities of government are vested in an external power and include not only colonies, but dominions, empires, protectorates, and the like. Republics, according to Plato's ideal, are those states where the government is by the people and for the people (Rowe, 1984). Many states now style themselves as republics to give notice that they are signing up to those ideals, even though this may be far removed from the reality of their governmental systems.

Historically, the distribution of these three categories has altered dramatically, especially in the past century (Christopher, 1999). In 1900, monarchies covered virtually the whole of Europe and Asia and republics were almost entirely confined to North and South America. In Europe, France and Switzerland were the only republics, and in the whole of the rest of the world there were just two others, the Orange Free State and the Transvaal in southern Africa.

In the year 2000 the picture could not be more different. Republics cover the bulk of four of the five inhabited continents, and the greater part of the rest comprise republics in all but name. In North America, Canada retains residual and largely token links with the UK through retaining the British monarch as head of state, as do Australia and New Zealand. For this reason, they are still technically classified as dominions. Monarchies with quite limited powers are still to be found in Europe, notably in Denmark, Norway, Sweden, the Netherlands, Belgium, Spain, and the UK. The emperor is still the head of state in Japan. There are also a few dynastic monarchies in the Middle East, the most significant being Saudi Arabia, and in south-east Asia, Thailand, Cambodia, and Borneo still have monarchs in the form of a king, a prince and a sultan respectively. In Africa, the King of Morocco is the only true remaining monarch, although there are monarchs too within South Africa, in Swaziland and Lesotho.

The transformation is important for what it says about the nature of the world map. Across the globe peoples have begun to take control of their own destinies, defining their identities in terms of how they are governed. The process is far from complete and for most people participatory democracy is a very distant prospect, but the change of attitude as to what constitutes good government is fundamental: even where monarchies survive they do so only with popular approval and true colonialism still remains in just a few small and scattered territories, many of them island dependencies.

Key themes and further reading

This chapter is about states and how they evolved as a framework for government. The role of capitalism and the attendant industrialisation and urbanisation on the development of states should now be appreciated, as should the historical significance of nationalism. The impact of IT and other technologies on the power of states should be clear. Historically, the spread of the state idea across the globe is an important theme, as is the different kinds of state governance, ranging from monarchies to republics, with colonial dependencies in-between.

There is a huge literature on the nature of states and on nationalism written by political scientists. Political geography has taken a great deal from this work and it has provided a most useful background for geographical writing on the subject. R. J. Johnston (1982) in *Geography and the State* has provided a readable and concise introduction to the geography of states, but those wanting to find out how other social scientists have approached the subject will find Ernst Gellner's (1983) *Nations and Nationalism* stimulating and challenging, particularly his discussion of nationalism. Karl Deutsch (1978) is a political scientist whose writing is readily accessible and in *The Analysis of International Relations* he provides a useful history of the spread of states since the middle of the eighteenth century. An up-to-date geographical survey of the great variety of states in the world and how they have evolved since the turn of the twentieth century is provided by A. J. Christopher (1999) in his *The Atlas of States*.

4 Making states work

The variety of local state systems

They that are discontented under *monarchy*, call it *tyranny*; and they that are displeased with *aristocracy* call it *oligarchy*: so also, they which find themselves grieved under a *democracy*, call it *anarchy*, which signifies the want of government; and yet I think no man believes, that want of government, is any new kind of government.

(Thomas Hobbes, *Leviathan*, 1651, part ii, chapter 19)

The infrastructure of the local state

The national area of all states is politically subdivided for one reason or another, be it administrative convenience, ideological reasons associated with a desire to promote local democracy, ease of economic management, or any of a host of other reasons. Yet, whatever the logic, as Thomas Hobbes pointed out so acerbically over three centuries ago, the people who are governed always tend to feel let down by those in power who rule their lives. As a result, there is almost always an ongoing process of change, in that changing political, economic, and social circumstances demand that the infrastructure of the local state be continually updated and revised. Against this inbuilt dynamism, the constitutional terms under which the local state was initially constructed, popular attachment to existing structures, and vested political interests combine to bolster the status quo, slowing down, and even denying, the inevitable necessity to adapt (Kumar, 1988). In some countries adaptation has been almost continuous. Romania, for example, underwent a dozen different administrative reforms in the first half-century after its foundation as a modern state in 1919, reflecting the turbulent political history in the period (Helin, 1967). Elsewhere, as in the USA for example, change has been much more evolutionary, with basic federal administrative divisions being slow to alter, except to accommodate shifts in the distribution of population and new states joining the union (Brunn, 1974).

The most stable local state systems are those formally specified in a constitution. In the US Constitution or the German *Grundgesetz* the structure of local government within the federal state is clearly defined, along with the separation of powers between the different levels of government within the federation. Foreign policy, overall economic management, and the highest courts are almost always matters for the central state, but matters such as education and the provision of many other services are usually the statutory responsibility of the lower levels of government. The central state can only intervene in these matters when a dispute arises about the competence of a particular level to act in given circumstances.

At the other end of the spectrum are unitary states, where the central government retains total control over all aspects of administration and policy. In other words, although they may be subdivided into local state systems, any powers are only delegated on sufferance by the central state and may be rescinded. Between these two extremes are many variants, with different degrees of delegation, often justified under the banner of devolution and bringing power closer to the people.

Nor is the process necessarily one way. A popular description for the recent evolution of state systems, particularly in Europe, has been the so-called 'hollowing out of the state' (Jessop, 1994). With the growing interconnectedness and globalisation of all aspects of society since the second half of the twentieth century, both public and private international institutions have become increasingly important. Political bodies such as the EU have encouraged functions previously controlled by states to be ceded to these new bodies. This in turn has led to demands for sub-national entities to be given greater autonomy and, in some instances, to bypass the national state apparatus in favour of the new supra-national bodies. A notable instance of such realignment is the regional policies of the EU. There has been considerable friction between some member states, including the UK, at the way in which the European Commission has taken the lead in deciding which sub-national regions are most deserving of assistance, with little if any reference to national governments (Keeble, 1989; European Commission, 2001).

It should also be pointed out that territorial separation of powers is only one aspect of the structure of governance. The issue of the separation of the powers of church and state, and of the distinction between the sovereign, the government, and the judiciary are all at the heart of understanding state power. These overlapping powers frequently lead to territorial inconsistency, with reform by one area not being matched by others. For the most part these inconsistencies tend to be relatively minor, but there are instances where the judicial or religious administrative maps are markedly out of kilter with state administrative subdivisions, often reflecting a bygone age.

Napoleonic Code

One of the most ambitious and successful attempts at specifying the nature of civil society and the division of responsibilities and power within a state has been the Napoleonic Code, devised by Napoleon Bonaparte in France in the wake of the French Revolution at the beginning of the nineteenth century. It guaranteed equal rights for all (though, significantly, women were not initially included in the so-called universal franchise and did not have a vote); it abolished privileges of birth and the widespread practice of serfdom; it separated church and state; it allowed everyone to work in the occupation of their choice; and it strengthened the status of the family, including clarifying and defining the laws of inheritance. Under separate legislation, Napoleon also completely reformed the administrative and judicial systems of France, setting up the network of *départements* that today still form the core of the local government system in France.

Nevertheless, even this thoroughgoing revision of political and social structures did not completely succeed in obliterating the pre-revolutionary map. As can be seen from Figure 4.1, the civil political map of France is substantially different from the map of judicial jurisdictions, which is still closely based on the earlier subdivisions. In some respects this is nothing more than a curiosity, but it does lead to annoying anomalies in terms of trying to coordinate legal and other administrative functions. It is, however, a striking illustration of the separation of the government and the judiciary, which was one of the key goals of the Napoleonic reforms.

The influence of the Code was enormous, reaching well beyond the boundaries of France itself. It was introduced across the whole of the Napoleonic Empire in both Europe and overseas (Figure 4.2) and its basic structure forms the core of the civil systems in thirty-one

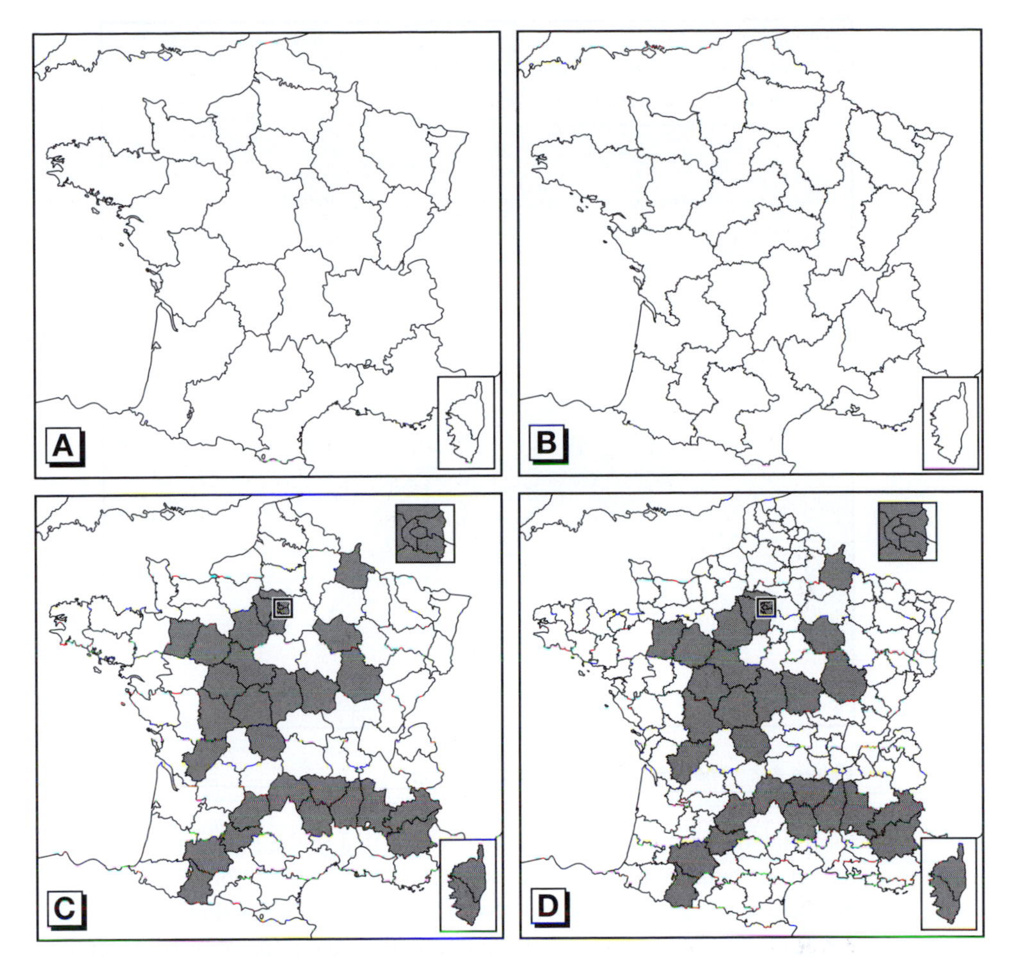

Figure 4.1 ***The administrative and legal map of France*****:** ***A administrative*** **régions;** ***B appeal courts*****;** ***C administrative*** **départements;** ***D high courts***

present-day states in Europe and Latin America. It has been every bit as important as the US Constitution in reshaping the political map in large parts of the world.

The state apparatus

The institutions of local government, together with the mechanisms of political representation and intervention, are collectively described as 'the state apparatus' (Clark and Dear, 1984). As a set of structures, they cannot of themselves exercise direct power, but inevitably they are

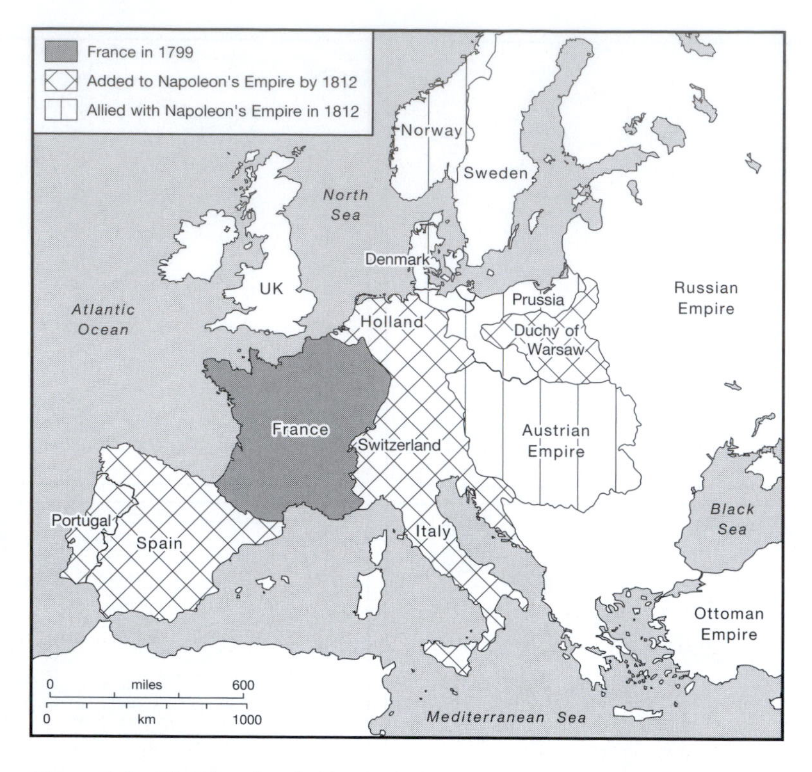

Figure 4.2 ***The Napoleonic Empire in Europe***

formulated and manipulated in such a way that in their operation they reflect the balance of power in society and reinforce the prevailing political ideology. At the most obvious level, this means that those in government will always seek to fine tune the state apparatus, by repealing elements likely to work against their policies and by inserting new ones that further their aims. However, because the mechanisms of change are usually cumbersome and tend to work relatively slowly, the state apparatus is also geared in part to historical aspirations and visions of the nature of the local state.

Given the complexity of modern society and the scale of state activities, which normally cover large portions of the social security system, education, and policing, to name but a few of the areas that come within its purview, the state apparatus spans a very large number of different purposes and for this reason is difficult to summarise simply. Nevertheless, Figure 4.3 describes the main facets, dividing the functions into four broad areas of sub-apparatus under the headings, consensus, production, integration, and executive.

	Functions			
	Type I Consensus	Type II Production	Type III Integration	Type IV Executive
Sub-apparatus	political	public production	health, education and welfare	administration
	legal	public provision	information	regulatory agencies
	repressive	treasury	communications and media	

Figure 4.3 ***The state apparatus***

Source: after G. L. Clark and M. Dear (1984) *State Apparatus*. Allen & Unwin, Boston MA.

Consensus

The consensus sub-apparatus exists to ensure that as far as possible all sections of the population are included in the social contract between the people and the state, including acceptance of the mechanisms by which the contract is enforced. It usually involves three distinct elements. The *political institutions* are devices that enable individuals to express their views formally about those who seek to govern them and include provision for elections, the right to establish political parties, and formal constitutional arrangements. The *legal institutions*, which include the formal legal statutes and the whole range of court structures, enable disputes of all kinds to be settled within an agreed framework. These disputes may be between individuals and groups and the state, between individual citizens, or in some jurisdictions between individuals and the church. The *repressive institutions* are the medium of control, such as the civilian police and the military, backed up by prison and probation services.

The balance between political, legal, and repressive institutions varies depending on the nature of the state. In democracies, the form of government favoured by international bodies such as the United Nations and firmly established in North America, most of Europe, and the Indian subcontinent, the prime purpose of the political institutions is to ensure that the government is representative of an electorate consisting of the entire adult population, with votes all counting equally and citizens allowed to cast them freely for any opinion without intimidation by the state apparatus (see Chapter 7). In practice, it is an ideal that is very hard to achieve and sustain fully. There are always limits imposed on who can or cannot vote, even in states with so-called universal suffrage. Normally, the right to vote is restricted to 'adults', but at what age people graduate to adulthood is subject to continual downward revision, standing currently in most cases at age 18. The terms under which people qualify as citizens with the right to vote also varies between states and there are always sections of the resident population, usually recent immigrants, who do not qualify.

The legal institutions, which include the courts and judiciary, should be completely independent of the government in a democracy, so that they can operate dispassionately in disputes and prosecutions. In practice, such a complete separation is hard to sustain, even in the most advanced and well-established democracies, if only because heads of state are invariably involved in judicial appointments. Where a democratic framework is absent, or poorly embedded, as is the case in large parts of the world, there is often not even the pretence of a separation. Accusations that legal institutions were little more than the agents of government were one of the fundamental objections that democratic countries levelled against the former Communist states in the Soviet Union and Europe.

Repressive institutions are a necessary element of all societies, but the problem for states is how to ensure that bodies, like the police and the armed forces, do not abuse the considerable powers vested in them. It is in this context that the independence of the legal institutions is most severely tested and all states must be vigilant against abuses of state power. There is constant scrutiny: from the United Nations and other international bodies, from other states, and from non-governmental organisations, such as Amnesty (see 'Human rights' in Chapter 12). Together they provide a powerful, though by no means totally effective, check on the misuse of state power.

Production

The production sub-apparatus exists to preserve and nurture the preconditions for capitalist economic accumulation (see 'The concept of the state' in Chapter 3). This involves seeing that structures are in place to ensure that the economic system remains in good health, though a significant part of the role is undertaken elsewhere in the form of measures aimed at ensuring the well-being of the population and the reproduction of the labour force. *Public production* incorporates the whole range of state-produced and distributed public goods and services, such as public roads and other significant parts of the transportation system, all of which are difficult to offer entirely through private suppliers. *Public provision* refers to goods and services produced on behalf of the state through private agencies, an obvious example being the infrastructure of public buildings erected by private construction companies through state-let contracts. *Treasury controls* encompass the whole range of fiscal and monetary polices employed by the state to regulate its internal and external economic relations. For example, many states operate strict controls over the movement of national currencies beyond the borders, seeing them as a way to maintain control over their national economies. In a world where the development of capitalism and free-market economies are widely viewed as the necessary prerequisites for economic growth such restrictions are an anathema, so that a great deal of international effort has been directed towards seeing such restrictions reduced, or eliminated (see Chapters 11 and 12).

Integration

The integration sub-apparatus is designed to foster the physical and social well-being of the population. Central to this is the gamut of *health, education and welfare services*, which also incidentally help to ensure the reproduction of the labour force. In a sense, as mentioned above, this could equally well be seen as part of the production sub-apparatus, though the population probably prefers to see it presented as part of the provision for social support. *Information services* are the mechanisms employed by the state to control the flow of information about its activities. There are very considerable differences between states in the degree to which they allow access to information about their activities, with some, such as the USA, guaranteeing freedom of information by statute, while others control any release in an entirely ad hoc and whimsical fashion. *Communications and media* are the instruments for controlling the dissemination of information. Most states have taken steps to ensure that there are nationwide radio and television services and national newspapers and many have gone further by promoting regional and local equivalents as well.

Executive

The executive sub-apparatus ensures that the other elements of the state apparatus – consensus, production, and integration – operate efficiently and in such a way as to reinforce one another. There is a high likelihood that if the range of state functions is left uncoordinated they will actually work against each other and undermine, rather than support, the common purpose. For this reason, *administration* has become one of the key arms of government. This in turn has increasingly spawned *regulatory agencies* to monitor performance. At all levels audit is now a crucial function and audit offices have an extremely influential role in judging the performance of government and its agents and in insisting that it lives up to agreed standards.

Governance or government

The kind of rigid structure set out in the previous section has been subjected to growing questioning in recent years, a debate encapsulated in the meaning of the word 'governance' (Painter, 1995). The traditional definition defines governance simply as the act of governing, and thus as synonymous with the word 'government', but it is now much more usually taken to incorporate a wide range of institutions and organisations, both governmental and non-governmental, involved in the process of governing, and the relationships between them. Jessop (1997) defines governance as 'the "self organisation" of inter-organisational relations', while Rhodes (1997) refers to 'self-organising inter-organisational networks'.

The crucial point is that use of the word governance recognises the complexity and fluidity of the relationships between the whole gamut of organisations involved in government and the inappropriateness of the hierarchical and highly structured view of government. Rhodes (1997), who did much to develop a general theory of governance, talks about there being a policy network and argues that this ought to be the basis for analysing governmental systems. He identifies four key features common to all such policy networks (see Figure 4.4):

1 interdependence between organisations
2 continuing interaction between network members
3 game-like interactions
4 a significant degree of autonomy from the state.

Type of network	Characteristics of network
Policy community/ territorial community	Stability, highly restricted membership, vertical interdependence, limited horizontal articulation
Professional network	Stability, highly restricted membership, vertical interdependence, limited horizontal articulation, serves interests of profession
Intergovernmental network	Limited membership, limited vertical interdependence, extensive horizontal articulation
Issue network	Unstable, large number of members, limited vertical interdependence

Figure 4.4 ***A typology of policy networks***

Source: after R. Rhodes (1997) *Understanding Governance*. Open University Press, Buckingham, p. 38. Reproduced with permission of the Open University Press/McGraw-Hill Publishing Company.

In other words, those organisations actively engaged in the process of government must have clearly defined and interdependent roles. The interactions between them must be ongoing and meaningful for the whole process of government. To a degree, the interactions must also be formalised, in the sense that the organisations involved must accept their own and each other's roles and operate within that set framework. Finally, they must, to a significant degree, operate on their own terms and not simply as agents of the state, acting as some omnipotent higher authority. In such a scenario, the state becomes just one element of many, albeit a very important one, in the structure of governance (Jones, 1997).

The implications of this reading of the way in which states are organised and operate internally are considerable. The loss of a fixed hierarchy with the central state government at its head means that new hierarchies can emerge through a process of negotiation or the exercise of power, thus introducing fluidity and variety in place of an established order. All kinds of groups can seek to dominate the local political process, with the ever-present danger that they may subvert the democratic process.

The rise of Militant

A potent example of the way a small and unrepresentative subgroup may succeed in dominating government and the political process is the rise of Militant in the city of Liverpool in the 1980s. The Militant Tendency, to give it its full name, was a far left, Trotskyist faction within the Labour

Party in the UK, which systematically set about infiltrating and dominating the political institutions of local government in selected towns and cities. The so-called strategy of entrism (Taaffe, 1995) was devised in the late 1950s and then over a period of 25 years was implemented with growing vigour. The hallmark of entrism is the subversion of an open and established political party or pressure group to pursue an agenda different from that officially agreed and this was at the root of the growing unease within the Labour Party at the burgeoning influence of Militant.

In governmental terms, Militant's greatest success was in the city of Liverpool. Under the leadership of Derek Hatton the faction came to dominate the city council and from this power base mounted a serious challenge to the authority of the enfeebled Labour Party in the late 1970s and early 1980s. Initial attempts by the national party to reassert its authority were notably unsuccessful and Militant was further emboldened by its involvement in a number of high-profile labour disputes, notably the miners' strike in 1984. In 1985, however, the Labour Party at its annual conference mounted a far stiffer challenge and Militant's policies and bid for even greater official recognition were roundly defeated and the organisation became proscribed.

Militant never recovered and its influence rapidly declined, to a point in 1994 where it was formally disbanded as a separate group. It lived on as a minor element on the socialist fringe as Militant Labour, before renaming itself in England as the Socialist Party, but it has never regained any significant political power base. In Scotland it did enjoy somewhat greater success as Scottish Militant Labour, occasionally successfully challenging the Labour Party at the ballot box, without ever posing any serious threat to its authority.

Subsidiarity, devolution, and regionalism

The concept of subsidiarity

Subsidiarity is an important principle underpinning the whole concept of the local state (Box 4.1). The idea that decisions should be taken as close as possible to those directly affected is central to democracy and, implicitly, argues strongly in favour of local control and devolved and decentralised government. It is particularly important with reference to

Box 4.1

Subsidiarity

The principle of subsidiarity requires that decisions in the EU are always devolved to the most appropriate local level. It was first included in the Treaty on European Union signed at Maastricht in 1992 and applies generally to decisions taken within the EU, but with two important exceptions. First, the principle of subsidiarity is not retrospective, so that decisions taken before the treaty came into force are not affected. Second, it does not apply to areas of decision-making where authority is specifically vested in the EU Commission, such as environmental quality standards. In practice, the principle of subsidiarity has proved difficult to operate, partly because of these exceptions, and partly because there is no clear definition of what constitutes the appropriate level to which a given decision should be devolved.

the EU and the growing number of states that are now members. The EU has always been committed in principle to subsidiarity, but since 1986 has been formally committed by treaty to achieving it. By the middle of the twenty-first century the EU will encompass a significant proportion of the world's independent states and of its total population, making its commitment to subsidiarity very significant (Inman and Rubinfeld, 1998). At the inception of what is now known as the EU in 1952, it embraced six countries with a total population of about 200 million. By 2004 this had grown to twenty-five countries with a population in excess of 450 million, with the prospect of as many as a further five countries joining in the foreseeable future. This could increase the total population of the EU to over 600 million, or roughly 10 per cent of the world population. For all these countries and the people in Europe, therefore, the commitment to the principle of subsidiarity will ensure that the local state and local decision-making are going to be increasingly important in determining how they are governed.

More generally, a commitment to subsidiarity almost inevitably has a direct effect on the nature of local government, with increased emphasis on sub-national political structures and a predisposition to see power delegated to them. In many countries, the recognition of regional identities has become a powerful political issue within the framework of the nation state, leading to insistent demands for greater political autonomy at a regional level.

In countries such as France, there is a long tradition of regionalism and the regional tier of government is well established. There are twenty-one *régions* in European mainland France (including Corsica) and they have played a significant role in promoting distinctive regional identities. In Germany, the fifteen *Länder* and Berlin are at the core of a regional structure built into the Constitution (*Grundgesetz*), recognising in a formal way the wide variety of different political traditions and histories that make up the present-day Federal Republic.

Elsewhere, the growing demands for greater regional autonomy often reverse a trend towards central state dominance in the early part of the twentieth century. This has been particularly marked in Spain since the end of the Fascist dictatorship of General Franco in 1975. Always an uneasy amalgam of different regional groupings, many of the Spanish regions have become much more assertive and aggressive in arguing for greater independence from the national government in Madrid. The Basque separatists in north-eastern Spain have been waging their campaign through unashamedly bloody terrorist policies ever since the end of the Spanish Civil War in 1939, but since Spain joined the EU in 1986, they have turned more, though by no means exclusively, to democratic methods in arguing their case. The Basques have undoubtedly been encouraged to review their tactics by the success of the Catalans in the north-east of the country in achieving substantial independence for Catalonia through democratic means.

Devolution and the remaking of the United Kingdom

The name of the United Kingdom of Great Britain and Northern Ireland is something of a hostage to fortune in that it does somewhat dangerously flaunt the unity of what is manifestly an amalgam of distinctive, if not disparate, regions – England, Northern Ireland, Scotland, and Wales (Nairn, 1981). In its present form, the UK dates from 1921, with the partition of Ireland and the creation of a semi-autonomous region in Northern Ireland, but there have been a number of significant realignments within that structure over the intervening years. In 1972 the devolved administration in Northern Ireland collapsed in the face of growing turbulence and dissatisfaction on the part of the minority Catholic population with the impenetrable political dominance of the Protestant majority. Since then there have been repeated attempts to reinstate a devolved government, notably in 1973 with the ill-fated Power-Sharing Executive, and, after a decade of painstaking negotiation, in 1998 following the Good Friday Agreement. It was all to no avail and the devolved assembly was again suspended and direct rule from England reinstated.

Elsewhere in the UK, and very much in line with the strenuous efforts to find an acceptable form of devolved government for Northern Ireland, calls for devolution in Scotland and Wales were becoming increasingly loud. After referenda in both 'countries', in 1998 a full-scale parliament was elected in Scotland and a rather more limited assembly in Wales. In both cases one of the main arguments deployed in favour of devolution was that there were already countries in the EU, notably Luxembourg, that were full voting members with much smaller populations and economies than either Scotland or Wales. The situation was also set to become even more anomalous in the eyes of the Scots and the Welsh when the EU expanded to twenty-five members in 2004. Countries such as Malta, Latvia, and Estonia, all small independent states, were then set to become full members.

Once the principle of devolution had been ceded in Scotland and Wales, other less clearly distinct parts of the UK began to press their claims for greater independence from central government and formal recognition of their regional identities (Jones and McLeod, 2004). Regional devolution in England has been hotly debated in recent years and more and more administrative functions have been relocated to the nine newly created Regional Development Agencies. Unelected Regional Assemblies also provide for some limited discussion of future policy in the individual regions by groups of key decision-makers, a process reminiscent of the early years of the EU itself, when an appointed parliament was the only mechanism catering for an input to its deliberations, other than through the national governments of the member states (Blacksell, 1981). Whether the process will go even further and move to elected regional government is open to question. Such a change was certainly intended by the central government but, as the first referendum on whether or not to set up an elected Regional Assembly in the north-east of England decisively rejected the idea, it is unclear if it will be pursued further for the time being.

Devolution in Scotland and Wales has already significantly altered the balance of political power in the UK and, if Regional Assemblies do eventually take root in the English regions as well, then the shape of the UK within the EU will be radically changed. What is more, an enthusiasm for devolution could well spread to other large and highly centralised states in the EU, changing the shape of their domestic governance as well.

Regionalism and economic management

Much of what passes for regionalism in the literature is more accurately described as regional economic development and the need to counteract the effects of uneven development (Massey, 1979 and 1984; Lee and Wills, 1997). There is a widespread appreciation that inordinate economic disparities within states will inevitably lead to political discontent and instability and that regional policies in some form are inevitably part of the process of government at the sub-national level (Friedmann and Alonso, 1964). Indeed, in the middle part of the twentieth century, the identification and development of economic growth poles was seen as something of a panacea for promoting orderly economic progress (Boudeville, 1966).

Regional policy can follow a variety of paths and, in practice, almost always incorporates a number of different elements. The most widely used set of instruments are those that attempt to modify market conditions in favour of a particular region. In some form or other, all governments offer inducements to potential investors, such as tax breaks, the waiving of site rents, or direct cash grants. The use of such inducements is so common amongst government at all levels that it is difficult to pick out any of particular note, but the identification by the EU of economically deprived areas across member states (the Objective 1 and Objective 2 areas) are good current examples (Figure 4.5 and Box 4.2).

Maintaining and improving the general physical infrastructure through the provision of roads, railways, and all kinds of communications facilities, as well as providing housing and a high level of environmental quality is another universally used strategy. A particularly striking example of such state investment is the huge infrastructural investment in the former East Germany by the Federal Republic after unification in 1990, conservatively estimated at over €1 trillion. A significant part of this investment has been in new and improved telecommunications and road and rail links, such as the extension of the intercity ICE express train network into the new *Länder* and the rapid expansion of the motorway system, so that the whole of the unified Germany has a more realistic chance of operating as an integrated economic area.

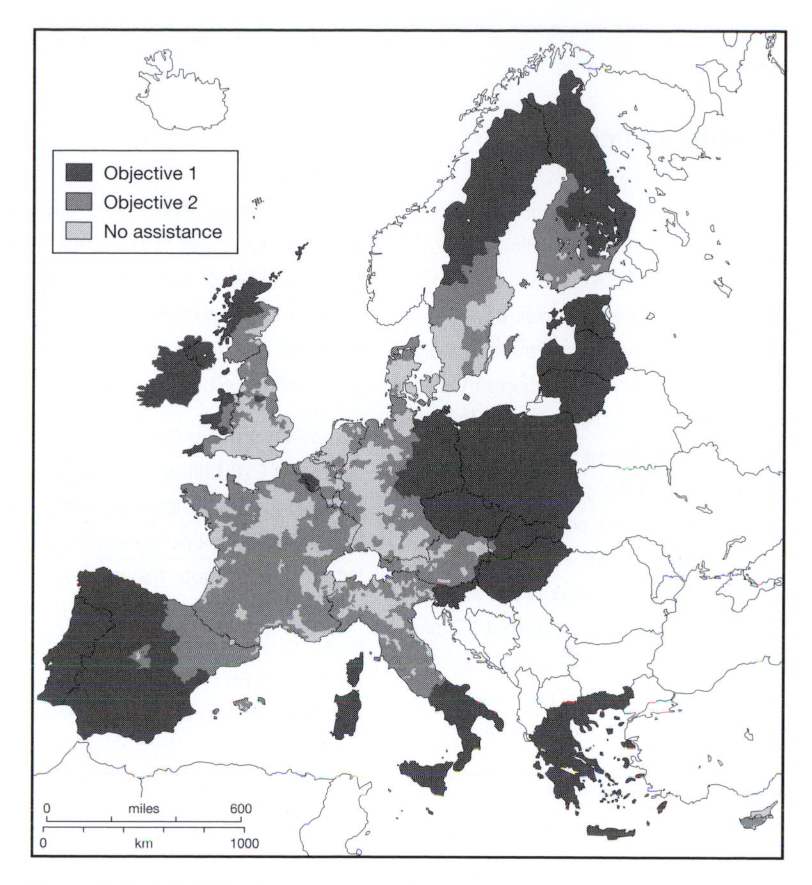

Figure 4.5 ***EU Objective 1 and 2 regional assistance areas, 2004***

Governments also frequently take more direct measures to redistribute regional economic activity, by directing government offices to specific locations, or by awarding government contracts to contractors where they want to stimulate economic activity. The latter applies particularly to defence contracts and in the USA an enormous amount of political lobbying by members of the Senate and Congress is devoted to securing government investment for their own particular state. Broadly the process is self-cancelling, but there are variations from the norm and some of the smaller and poorer states, such as West Virginia, are heavily dependent on federal defence contracts.

Regional policies are rarely driven purely by economic or narrow political considerations, and social welfare almost always plays a part as well. Landscape, scientific, and heritage designations are primarily

Box 4.2

EU regional policies

The bulk of the money for regional development within the EU is allocated to the Structural Fund. In 2004 the amount available under this heading was €195 billion. The largest element of the Structural Fund is Objective 1, which supports those regions that are economically lagging behind the EU average. Objective 1 regions, which cover 22 per cent of the population of the EU, are allocated 70 per cent of the Structural Fund. Objective 2 regions, which cover 18 per cent of the population and are also lagging economically, receive 11.5 per cent of the money. In these areas it is judged that modest support for industrial restructuring is likely to improve their relative position. Some 12.3 per cent of the money is allocated to Objective 3, the purpose of which is to stimulate new employment anywhere in the EU, except in Objective 1 regions where this goal is already covered.

The much smaller Cohesion Fund was allocated €18 billion in 2004. It comprises four programmes: Interreg III to promote transnational cooperation within the EU; Urban II to promote the development of deprived urban areas; Leader to encourage local development initiatives across whole areas; and Equal to combat all kinds of discrimination in the labour market.

categorised for their own sake, rather than in the pursuit of any direct economic enhancement. In the UK, designations such as national parks and sites of special scientific interest (SSSIs) reflect a commitment on the part of government, born out of steady public pressure in the early part of the twentieth century, to implement the vision of permanent protection for key national landscapes (Sheail, 1981; MacEwan and MacEwan, 1987). The result has only had a marginal and incidental influence on the economic map of the UK, but a profound impact on public perceptions of the land and its worth.

A passing thought

Memories of historical stability in the local, sub-national political map have almost certainly always tended towards exaggeration; constant political change at all levels is a fact of life. Nevertheless, the pace of change has undoubtedly quickened and as the fluidity of governance has come to be preferred to the more rigid concept of government, the local state has become less stable and more prone to failure. It is partly in recognition of this that the government in the UK established boundary commissions for England, Scotland, Wales, and Northern Ireland in 1986 to advise it on changes to electoral boundaries. The work of the commissions will be dealt with in detail in Chapter 7, but it is relevant at this point to acknowledge their contribution to maintaining the effectiveness of the local state in the UK.

Key themes and further reading

Understanding the internal workings and organisation of states and how they function – the local state – is central to a proper understanding of the dynamics of a significant part of political geography. Four operational subsets are at the heart of the local state: the consensus apparatus, the production apparatus, the integration apparatus, and the executive apparatus. The distinctions between them should be understood. The significance of the difference between governance and government needs to be appreciated. The concept of subsidiarity and its relationship to devolution is similarly important and leads on to the whole question of regionalism and equity in economic management. Finally, it must be appreciated that calls for stability and continuity in local state relations mask a reality of constant change.

An excellent introduction to the nature of governance and the working of the local state is provided by R. Rhodes (1997) in his book *Understanding Governance*. It explains in straightforward terms how states work and how decisions are made within local government. The economic and structural issues are discussed with force and clarity by Doreen Massey in two books, *The Spatial Division of Labour* (1984) and *Space, Place and Gender* (1995); the latter includes an important discussion of gender discrimination. A most readable account of the fragility of local state systems and the way in which they are continually reforming is given in Tom Nairn's case study, *The Break-up of Britain* (1981).

5 The politics of difference

> The conquest of the earth, which mostly means the taking it away from those who have a different complexion or slightly flatter noses than ourselves, is not a pretty thing when you look into it too much.
>
> (Joseph Conrad, *The Heart of Darkness*, 1906, chapter 1)

In many ways political geography has been turned on its head since the early 1990s. Influenced by a whole series of powerful writings in cultural geography (Harvey, 1989; Jackson, 1989), the focus has moved from order and certainty, and a search for clearly defined patterns, to the individuality of people and the multitude of continuously shifting relationships in which they are engaged. The acknowledgement of the importance of what is termed difference leads to the acceptance of a state of chronic instability in society, which is extremely difficult to accommodate within any ordered political framework. It is argued that difference is an all-pervasive feature of postmodern society, unsettling most apparent certainties and throwing into question any stable sense of identity (Jackson, 2000, p. 174). Philip Crang goes even further, arguing that the world of difference undermines any sense of authenticity, replacing it with a world of displacement and dislocation (Crang, P., 1996; Crang, M., 1998).

This postmodern reformulation is of crucial importance for two reasons. First, it breaks the apparently inseparable link between identity and place. People's sense of place is in a constant state of flux and has become increasingly global in its reach (Massey, 1995). Second, it removes any limits as to what constitutes a sense of place and a sense of identity. In the new world view, the ties binding people together are almost infinite in their variety and, as a result, are never something that can remotely be taken for granted.

Managing difference

Recognising difference and managing it are two very different things. The world may be complex, but the whole process of government demands it be simplified, even oversimplified, if it is to serve the needs of communities. Compromises are essential, even though they may be no more than temporary political expedients.

Language

Language, more than anything else, determines people's world view. Who people listen and speak to is crucial in shaping the way in which they think and act. At one level this is self-evident, but until relatively recently the role that language plays in political geography had not been fully appreciated. The Swedish geographer, Gunnar Olsson (writing in English of course!), has been pivotal in repairing the omission and raising awareness of what he terms 'linguistic geometry' (Olsson, 1980 and 1992). Previously, although the cultural significance of language had received wide, though intermittent, attention, and maps of language had been commonplace in the geographical literature, their cultural and political power had not been fully explored and put into context.

Subsequently, the situation has changed dramatically and geographers have focused extensively on language and the way it can shape the cultural and political landscape (Zelinsky and Williams, 1988; Jackson, 1989; Moseley and Asher, 1993; Doel, 1993). Nowhere is this more evident than in the gender bias built into most European languages, where traditionally discourses are almost invariably centred around male images. It is also proving very difficult to alter the focus to a more equitable gender balance (Bondi, 1997). National historical narratives are almost always formulated in terms of men and the way in which they have influenced the course of events, while the role of women is largely invisible, because the form of language used excludes a proper representation of their contribution. To a very large extent this reflects the realities of power within societies; women were almost entirely excluded from a role in formal political decision-making before the beginning of the twentieth century, and a hundred years later this is still the case in some countries in the world.

There are many instances where national governments have sought to guide and manage the evolution of language in order to strengthen the sense of national identity. A recent example of such direct intervention is the state of Israel, which adopted religious Hebrew as its national language when the state was founded in 1948. However, since Hebrew to all intents and purposes had only been used regularly for centuries as the language of worship, the world view it encapsulated was extremely narrow and its vocabulary had to be rapidly expanded for it to be usable as a modern means of communication. The efforts have met with considerable success and Hebrew is now the main spoken and written language of some 6 million Israelis, though it has not spread outside the country. In France there is a government commission, *La delegation générale à la langue française*, whose main duty is to review continuously the French language and ensure that it is not overly corrupted by imports from other languages.

One of the most determined attempts to develop and protect a national linguistic norm is the Gaeltacht in Ireland. As part of a drive to promote Irish Gaelic as a national language, an area of mainland and nearby islands on the west coast of Ireland, centred on the town of Galway, has been designated as the Gaeltacht. Its purpose is to act as a centre for promoting the Irish language and culture in the hope that the influence will spread to the whole of the country. However, despite the best efforts of the Department of Community, Rural and Gaeltacht Affairs, it has met with only limited success in preserving the active use of Irish Gaelic. Indeed, its main purpose now seems to be to promote tourism in western Ireland.

In Europe, the most vivid illustration of the hold that language can exert over political change is the debate that has raged in the EU about which national languages should, and should not, be officially accepted. The member countries of the EU have vigorously eschewed forming themselves into a federation and prefer to see themselves as, at most, a confederation with distinct national identities jealously preserved. As a result, there are twenty official languages, spoken and written translation facilities for which must be available at all times (Figure 5.1). Only five countries have not insisted on the inclusion of their own separate national language – Austria, Belgium, Cyprus, Ireland, and Luxembourg – though even some of them can request that documents be translated, notably the Irish requiring translation into Gaelic.

It is obviously a situation that is a practical impossibility as a permanent arrangement, especially as the EU is likely to expand further in the foreseeable future, but a solution is surprisingly difficult to find. English

Figure 5.1 ***Official languages in the EU***

is by far the most popular option as a common language, with 69 per cent preferring it, nearly twice as many as the second most popular alternative, French. Next comes German, though it is actually the most widely spoken as a mother tongue. On the other hand, less than 10 per cent speak Spanish as their mother tongue or as a foreign language, despite its widespread use elsewhere in the world. But the most surprising linguistic twist in the EU is the jump in the proportion of citizens able to speak Russian. Since the most recent expansion to twenty-five members in 2004, Russian has become the fourth most important language, though for obvious political reasons there is little chance of it being adopted as an official language in the EU.

Ethnicity and culture

Ethnicity and cultural difference have been amongst the most fraught and destabilising influences on the dynamics of the modern state and other political subdivisions. They challenge the whole idea of national norms and the belief in the efficacy of universal values, the very things that states strive to encourage in the search to foster national or regional political cohesion (Young, 1990). Attempts to celebrate and accommodate diversity are legion, but frequently they have foundered in the face of racial and cultural prejudice. All too often ethnicity is used as a loosely defined term to describe the habits and customs of minority groups in society, blithely ignoring the fact that majorities too have their distinctive customs and that being part of a minority is not a crime. More sinisterly, ethnicity is taken as being synonymous with the term race and used to demonise particular groups on the basis of their physical characteristics, even though the basis for such categorisations is extremely doubtful, let alone being morally justified (Mason, 1995).

Nevertheless, ethnic segregation is a fact of life in most societies, although the degree to which it is institutionalised varies enormously (Peach *et al.*, 1981; Peach, 1996). At one end of the spectrum it is a benign and even affectionate description, as for example with the use of the word 'Chinatown', but at the other it is used as an excuse for brutal and repressive segregation (Smith, 1989). At its most extreme, it takes the form of *de jure* policies, such as those of apartheid in South Africa, which restricted so-called coloureds, blacks, and whites to specific areas of the country (Smith, 1994 and Box 5.1). More frequently, though, it is a de facto process, whereby those who feel a cultural and ethnic affinity congregate together in specific areas of towns and cities.

In an ideal world, such differentiation should be of little political consequence, but in practice it leads to ethnic discrimination and persecution. It is an insidious process; what begins as apparently minor differences in levels of service provision can escalate quickly into overt repression, generating a violent response. In the UK, there was widespread horror at the urban riots that erupted amongst the black and Asian communities in the Brixton area of London in 1981, later spreading to other conurbations in the north of England. The subsequent report on the riots by Lord Scarman (1981) blamed much of the disturbance on the blatant and institutionalised discrimination against ethnic minorities. It made a whole series of recommendations with respect to policing and social provision to counter the discrimination, but

Box 5.1

Apartheid in South Africa

Racial segregation and white supremacy had been traditionally accepted in South Africa long before 1948, but when the Afrikaner Nationalist Party swept to power for the first time in the general elections held that year the formal policy of apartheid was born. Its purpose was separation of the races: whites from blacks, non-whites from each other, and one group of Bantu (the pejorative collective noun for all black Africans) from another. Those regarded as non-white included not only the majority black Africans, but also the so-called coloureds, people of mixed race, and people of Asian descent.

In the early years, the emphasis was on segregation in urban areas, which resulted in large parts of the coloured and Asian populations being forced to relocate away from white areas. Later, however, as towns and cities grew and began to engulf the black African townships, the black Africans too were forced to relocate to new, and more remote, townships. In the period between the Group Areas Acts of 1950 and 1986 it is estimated that some 1.5 million black Africans were forcibly driven from the cities into rural reservations. Even before the election in 1994 that brought Nelson Mandela and the African National Congress to power, the policy of apartheid was being dismantled as the Nationalist Party progressively lost its grip on political power.

these have proved extraordinarily difficult to implement effectively and ethnic discrimination remains an intractable urban problem.

Unchecked and in the absence of any democratic controls, the consequences of ethnic division can escalate into far more serious violence and what has come to be known as ethnic cleansing and genocide (Figure 5.2). Such state-sponsored crimes are horrific in both their intent and scale, let alone for the cold-bloodied way they are executed. In the former Yugoslavia during the 1990s, the Serbian army systematically eliminated both Croats and Bosnians from what they deemed to be their lands, using a five-stage strategy. They surrounded the area to be ethnically cleansed; they then executed as many of the community leaders and potential leaders as possible; they then separated out women, children, and old men; they then transported them to the nearest border; and they then executed all the remaining men (Donner, 1999). The pattern was repeated across southern Croatia and eastern Bosnia resulting in a whole series of notorious massacres, such as that

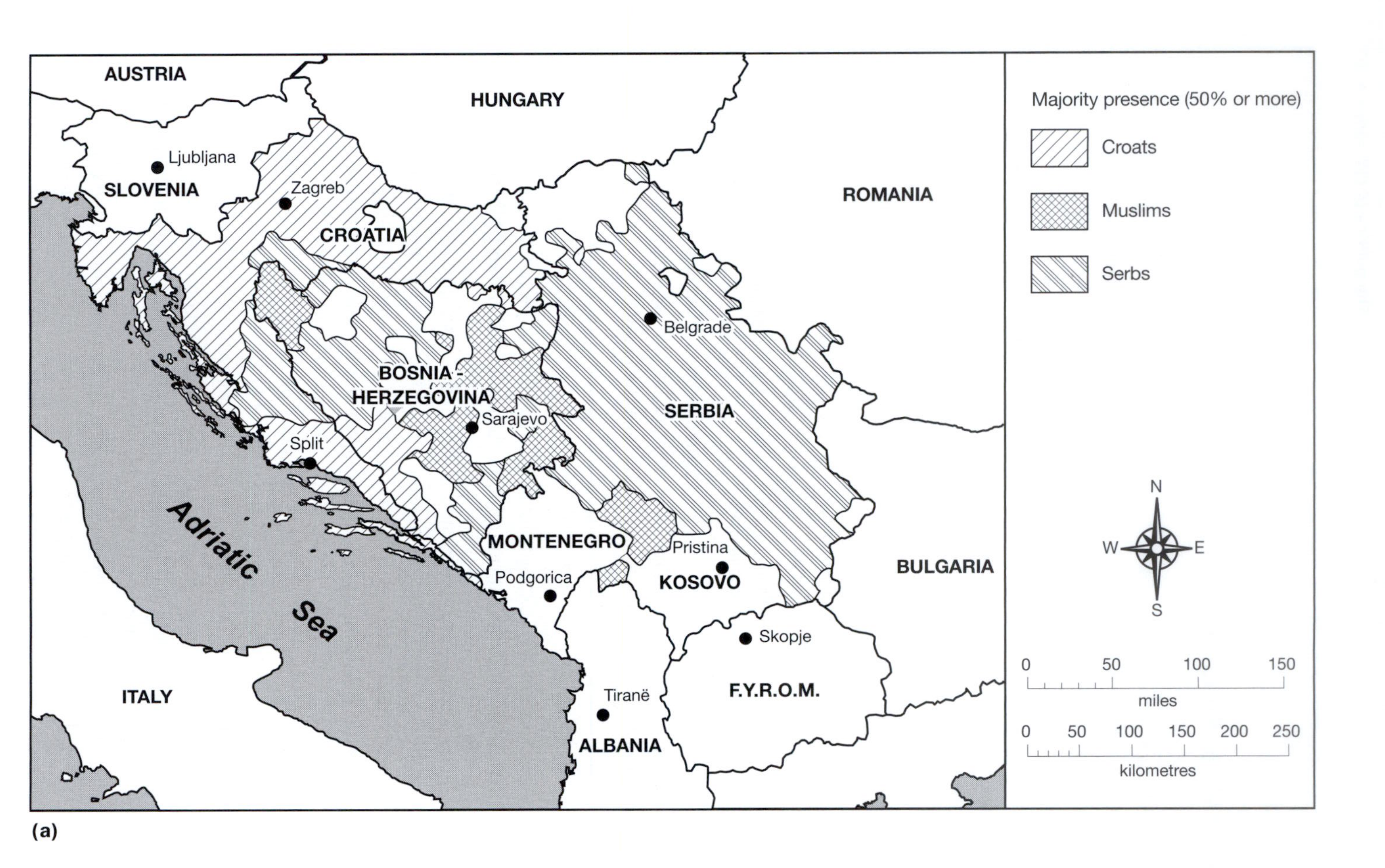

(a)

Figure 5.2 ***The process of ethnic cleansing in the former Yugoslavia*****:** ***(a) The mixed distribution of ethnic groups before the civil war; (b) The effect of the subsequent ethnic cleansing***

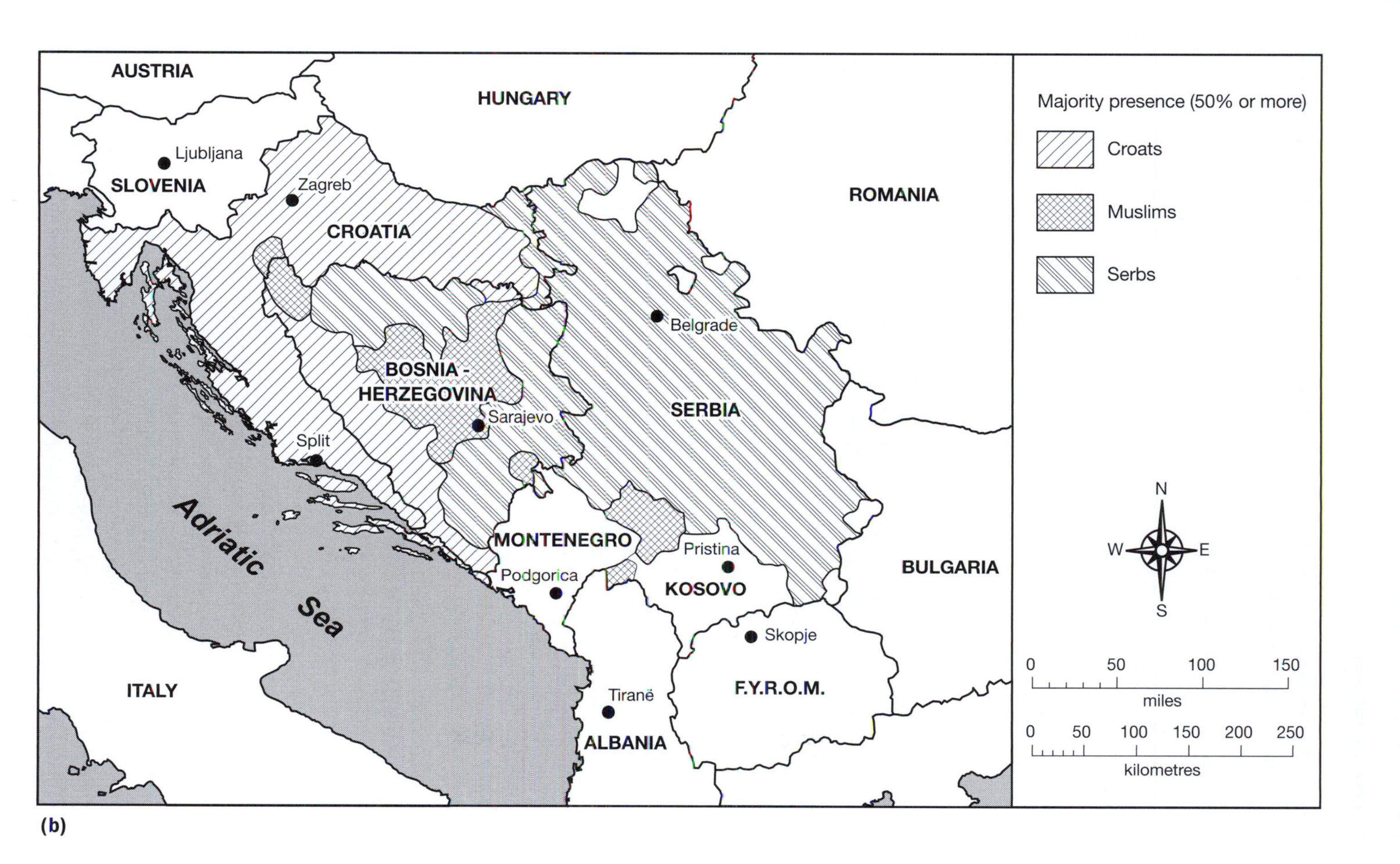
AUSTRIA
HUNGARY
ROMANIA
Ljubljana
SLOVENIA
Zagreb
CROATIA
Belgrade
BOSNIA - HERZEGOVINA
Sarajevo
SERBIA
Split
Adriatic Sea
MONTENEGRO
Pristina
BULGARIA
Podgorica
KOSOVO
Skopje
ITALY
F.Y.R.O.M.
Tiranë
ALBANIA
Majority presence (50% or more)
Croats
Muslims
Serbs
N
W
E
S
0
50
100
150
miles
0
50
100
150
200
250
kilometres
(b)

at Srebrenica in 1995 where 2,000 Muslim prisoners were executed in a single afternoon. There are other even more appalling examples across the globe, including the genocide in Rwanda in 1994, all with similarly brutal profiles (Box 5.2).

Box 5.2

Genocide in Rwanda

Between April and June 1994, it is estimated that about 800,000 Rwandans were killed in the space of 100 days. Two large rival tribes were at the heart of the massacre: most of the dead were Tutsis, and most of those doing the killing were Hutus. Violence has racked Rwanda frequently in the past, but even by its own standards the speed and scale of the slaughter devastated the whole population. The genocide was sparked by the death of the Hutu president, when his aeroplane was shot down on the orders, it emerged later, of the current Tutsi president, Paul Kagame, but it was just the touchstone for a particularly brutal and extensive flare-up in a long-running saga of violence between the Hutu majority and the Tutsi minority. The immediate violence came to an end when Tutsi rebels captured the capital Kagali, but over two million Hutus fled to neighbouring Zaire, creating a long-term refugee problem in the region. Although the new government has promised them a safe return, the legacy of the genocide remains. Over 500 people have already been sentenced to death, but there are about 100,000 more still in prison awaiting trial, while the UN International Court of Justice in the Hague struggles to find a way to bring the ringleaders before a tribunal to face charges of genocide.

Reading

Wood, 2001.

Religion

The most difficult of all differences to control within a political framework are those rooted in religion. Most states adopt a formal stance towards religious observance, be it explicit rejection as in China and other Communist states, a commitment to freedom of religious expression as in the USA and virtually all the Western democracies, or embracing a particular form as an official state religion, as with the Church of England, the Roman Catholic Church in Ireland, Italy, and Poland, and Islam across much of the Middle East. In the latter case, the state often underpins its chosen religion, either by ceding to it certain civil responsibilities, such as the Church of England's right to register marriages, or by encouraging and subsidising the construction of religious buildings, such as mosques in Saudi Arabia and many other of the Gulf States. There is, however, no part of the world where a single religion has a complete monopoly, so that tensions arising from different religious allegiances are a universal feature of society.

Very often these tensions coalesce into formal political parties, so that religious tension is almost institutionalised in the political process, a notable example being the Hindu Bharatiya Janata Party (BJP) in India, which is deeply distrusted by the Muslim minority. Indeed, in India, a religiously deeply divided country, one of the electoral attractions of the Congress Party that has formed the government for the greater part of the period since independence in 1947 is that it has no formal religious affiliation. In Northern Ireland, the Unionist parties almost exclusively represent the Protestant majority, while the Social Democratic Party and Sinn Féin represent the Catholic and Irish nationalist minority (Figure 5.3). Thirty years of conflict between the two factions mean that the division between them has become formalised on the ground in a series of religious ghettos, especially in the major cities of Belfast and Londonderry, or Derry as it is universally referred to by the Catholics (Boal, 1969). Non-religious parties, such as the Alliance, have little influence on the province's political landscape.

Religious buildings have been used quite explicitly on occasion to challenge the existing political order. Harvey (1979) argues persuasively that the Basilica of Sacré-Coeur in Paris was an overt attempt to help restore a more conservative political order in the wake of the insurrection of the Paris Commune in 1871. Although France became a republic after the Revolution of 1789, the Catholic Church exerted considerable influence on various French governments in the course of the nineteenth century and it was partly a reaction to this that sparked the urban rebellion of the Commune.

Nationalism and self-determination

Traditionally, in the modern era since the beginning of the industrial revolution in the early nineteenth century, the overt expression of political difference has been nationalism (see also 'The spread of states' in Chapter 3). It is a powerful concept that emerged first in Europe and

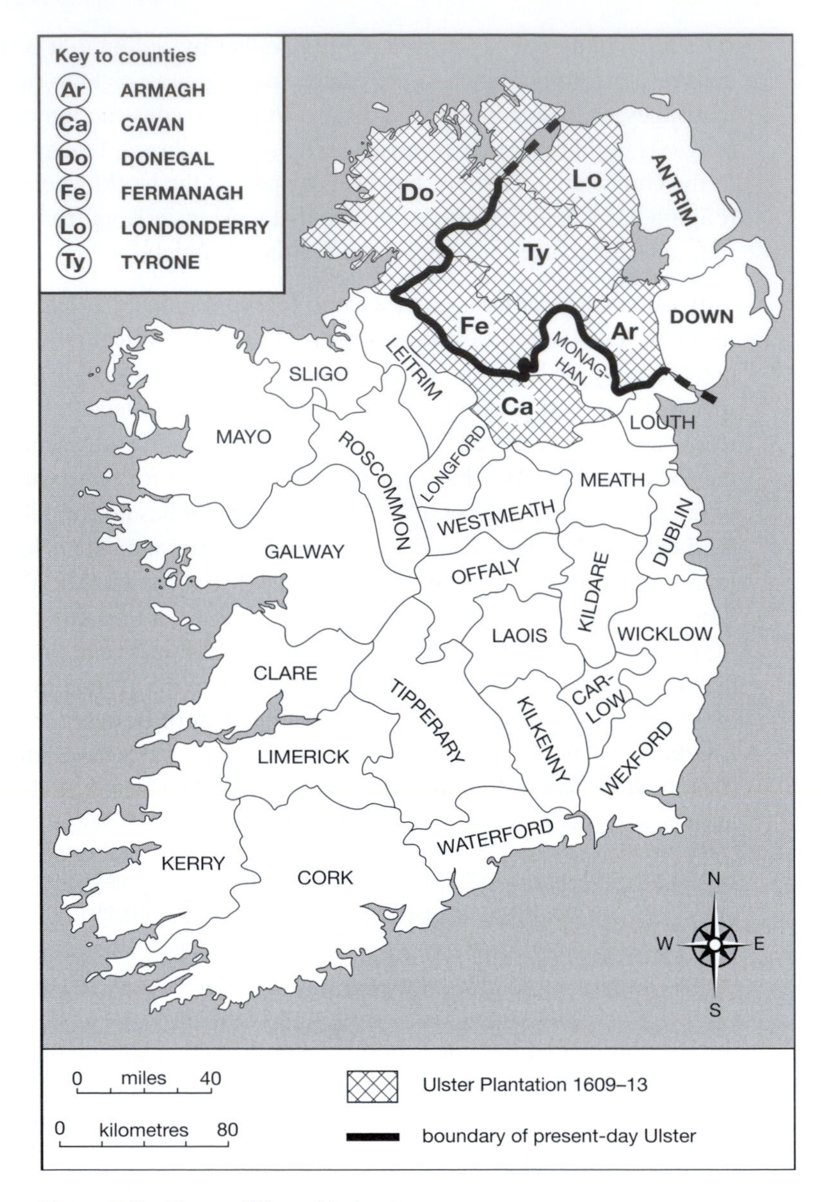

Figure 5.3 ***The partition of Ireland***

has subsequently spread right across the world. Its central aim is to achieve social justice through gaining sovereignty over an exclusive political homeland, the justification being that every nation has a natural right to self-government and control over its own affairs. Thus, whilst the politics of difference do not necessarily develop into a nationalist movement, this is in fact what has happened to many of the larger and better organised campaigns for the recognition of minority rights.

Almost by definition, nationalist movements in their early stages are a revolt against the established political order and, as such, are invariably seen as a threat and something that needs to be contained and even extinguished. As movements gain momentum, however, especially with the passage of time, they tend to gain legitimacy, but it is rarely easy for state governments to accede to secessionist, or irredentist, demands, because the dividing lines between the opposing groups are confused and overlapping (Chazan, 1991). There are always those who do not wish to secede and campaign vigorously for the status quo.

The Irish struggle for independence throughout the eighteenth and nineteenth centuries is a classic secessionist story (Davies, 1999). Born out of the brutal repression of Oliver Cromwell and the English Commonwealth in the mid-seventeenth century, a campaign spanning nearly two centuries culminated, first in the ill-fated Home Rule Act of 1914, the implementation of which was disastrously postponed because of the outbreak of the First World War, and finally in the Anglo-Irish Treaty of 1921. The Treaty led somewhat tortuously to the partition of Ireland, with the twenty-six counties in the south forming the overwhelmingly Catholic Irish Free State and then the Irish Republic, and the remaining six counties in the north and east forming the largely Protestant self-governing province of Ulster within the UK. The frontier between the two cut crudely between the two religious communities and the ensuing years have been spent trying to gain political acceptance for a very imperfect partition. The result has been an irredentist conflict, with the nationalist minority in Northern Ireland seeking to overturn the partition and the Unionist majority determined at almost any cost to resist a weakening of the province's position in the UK.

The origins of nationalism have been the subject of considerable debate (Gellner, 1983). It took root in Europe during the nineteenth century and there has been some argument as to whether industrialisation and modernisation have fed or starved its subsequent growth. Initially, it was thought that as societies developed and became more integrated, nationalism would gradually fade in the face of a more internationalised, or globalised, world but time has thrown increasing doubt on this interpretation. Nationalism has flourished and spread rapidly across the world, becoming particularly potent in the developing countries in Africa and Asia. As a result, it is now argued that it is not so much a result of industrialisation, as a reaction to its uneven spread. Many regions and countries have been left out of, or sold short by, the benefits of industrialisation and nationalism is now seen as a political response to try and reassert self-determination and control.

Nationalism is a broad and loosely defined term, incorporating a number of different movements. The most obvious is *state nationalism*, where governments appeal to feelings of national unity and promote symbols to bolster those feelings. National flags and national anthems, as well as swearing allegiance, are all universal examples of such manipulation In the USA, for example, the school day must begin by law with all students swearing their allegiance to the flag. Frequently, state nationalism is challenged by *sub-state nationalism*, when an irredentist group seeks to break away, or have greater independence from the central government. Examples are legion across the globe, but the long-standing campaign by the province of Quebec to loosen its ties to the federal government in Canada is a vivid case in point. In the developing world *anti-colonialist nationalism* is a very prevalent variant, whereby an indigenous people seeks to free itself from what it sees as a colonial oppressor. In the second half of the twentieth century such movements were especially active, resulting in political change and independence across virtually the whole of the Indian subcontinent and most of Africa. The biggest beneficiary, in terms of both area and population, was India which is now one of the fastest growing economies in the world. Finally, there is *ethnic and religious nationalism*, resulting from an oppressed ethnic or religious minority seeking to free itself from persecution and oppression. In most instances this can be seen as a variant of sub-state nationalism, but sometimes the persecution is so dispersed and virulent that a new state is established. As the USA spread westwards across North America, there were a number of examples, including Utah, which was founded as refuge for the Mormons, but, as with all the others, was eventually re-incorporated into the Union as a state in 1896. Elsewhere, there is the case of Israel, founded as a completely new state and Jewish homeland and committed to accepting Jewish immigrants from anywhere in the world without any restriction. Nationalist sentiment in Israel is extremely strong and proving very difficult to accommodate within the Middle East as a whole, provoking repeated intermittent conflict.

War and political breakdown

Any change to the established political order is unsettling and disruptive and often cannot be accommodated peacefully. When diplomacy fails the outcome is war and, recently, at any one time, there have been on average about forty different major conflicts across the globe. The seriousness and duration of wars varies enormously and most of the major conflicts are the result of unresolved power struggles between states rather than irredentism. Whatever their cause, they are a shifting and ever-present feature of the political landscape and the commonest mechanism of change to the world political map (Burghardt, 1972).

One of the main drivers behind the establishment of the United Nations Organisation in 1945 was to reduce the level of war in the world and to provide international peacekeeping forces to try to contain or prevent armed conflict (see Chapter 12). Since 1948 it has mounted fifty-nine such operations, though with the exception of the Korean War (1950–3), where at its peak there were over 900,000 troops under UN command, mostly from South Korea and the USA, the numbers involved in peacekeeping operations have been small. In 2004, there were sixteen operations on five continents, involving nearly 60,000 troops from ninety-seven countries, and for the most part they were undertaking policing, rather than combat, duties (Figure 5.4). In terms of service personnel, the most significant UN presence is in Africa, with troops in Sierra Leone, the Democratic Republic of Congo, Ethiopia and Eritrea, Liberia, the Ivory Coast, the Western Sahara, and Burundi, but it also stands between warring factions elsewhere across the globe.

In many instances the UN is not in a position to assist in preventing war breaking out, either because political disagreement within the membership ties its hands, or because it cannot finance a sufficiently large operation – or both. Most wars develop despite its best efforts and, once started, usually prove very difficult to end, not least because invariably what is really at issue is control of resources and any agreement on sharing is difficult, if not impossible, to achieve.

It is interstate wars that usually command the greatest public attention. In 2004, the most high profile was the invasion of Iraq by the USA, the UK, and to a much lesser extent other European and Asian countries. In little over a year the conflict claimed about 20,000 dead, 95 per cent of them Iraqis and it remains to be seen whether Iraq as a state in its present form can survive and, if not, what kind of new territorial order will

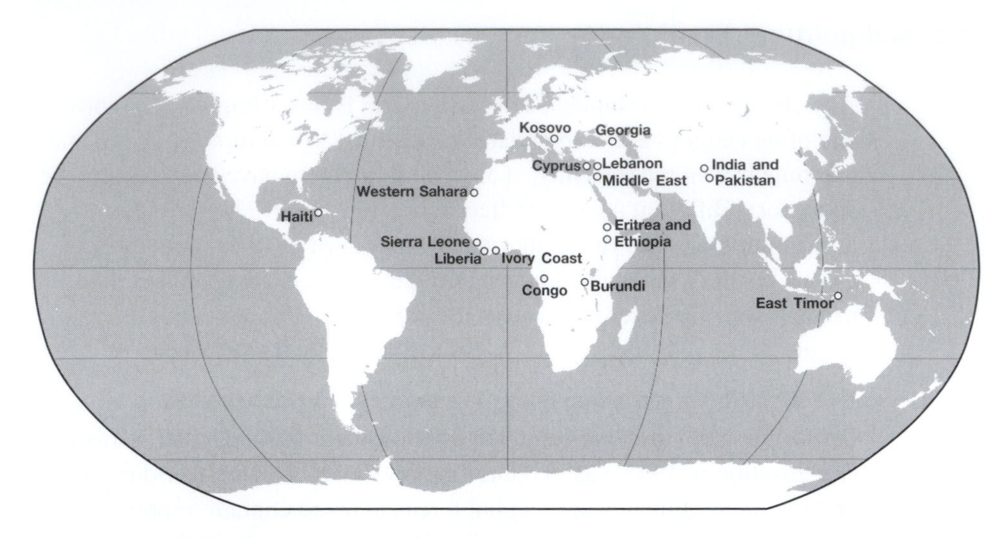

Figure 5.4 ***UN peacekeeping operations, 2004***

emerge. Iraq was previously governed by a despotic dictator, but it also had the second largest proven oil reserves in the world. Whilst overturning a tyrannical regime was undoubtedly a major factor behind the costly, crusading invasion, the promise of access to the oil was undoubtedly another, and it is now sustaining a determination to see through a very unpopular occupation, with massive geostrategic implications for the balance of global political power.

Elsewhere, a classic irredentist conflict over the future of Kashmir has spluttered on intermittently since 1947, threatening to involve the two regional powers, India and Pakistan, in face-to-face conflict. The dispute centres around whether Kashmir should remain a part of India, or become an independent state with closer links to the largely Muslim state of Pakistan. It has claimed over 65,000 Kashmiri lives and, even though peace talks between India and Pakistan started in 2003, it is still not clear whether they will be successful and what the future political status of Kashmir will be.

In Russia, the Republic of Chechnya has been waging a war to secede from the Russian Federation for more than a decade. Chechnya is a landlocked republic in the Caucasus Mountains between the Black and the Caspian seas and is seeking independence largely because its indigenous population is overwhelmingly Muslim. However, it is facing implacable opposition from the Federal Russian government, because it commands the most likely potential route for oil pipelines linking the

Caspian Sea basin with the Black Sea coast and, thus, access to the high seas. As a fully independent state, Chechnya would be in a position to dictate the flow of oil and Russia has been prepared to engage in a guerrilla war to safeguard its own interests. Whatever the politics, the conflict has resulted in at least 150,000 deaths since 1994 and shows no sign of being resolved.

Somewhat cynically, it may be said that such a pattern of international conflict represents 'normality'. There are always wars being waged; it is only the places where they erupt into general conflict that change. Actually, most wars are intra-state, civil wars between opposing factions within states and they too are most often waged over resources. In Colombia, for example, a conflict has been continuing for more than forty years between the government, aided by the US army, and at least four different rebel armies. At issue are drugs and control of the bulk of the illegal supply of cocaine to the world market. The USA justifies its support for the government by arguing that it has a legitimate interest in the drugs trade, as the major market for smuggled cocaine from Colombia. Its stance rather conveniently ignores the fact that all the major guerrilla groups – Fuerzas Armadas Revolutionarias de Colombia (FARC), Ejercito de Liberation National (ELN), Union Patricia Party (UP), and the Popular Army of Liberation (EPL) – are equally opposed to the drug trade in principle, but find the political oppression of the poor in both rural and urban areas by the government unacceptable.

In other instances, foreign intervention in civil wars is justified to prevent the conflicts from spreading to neighbouring states. This has happened often in Africa where the state and sub-state political frameworks usually date from the nineteenth- and twentieth-century European colonial era and, therefore, are at variance with the indigenous, pre-colonial, ethnic map. The civil war in the Democratic Republic of Congo between Laurent Desirée Kabila's Congo Armed Forces (FAC) and the Alliance of the Democratic Forces of Congo (AFDL) represents a recent manifestation of a conflict between two deeply opposed tribal groups, the Balendus and the Wahimas. It is a conflict that has involved a number of neighbouring states over the years, often because they too have large minorities belonging to these tribes. Namibia, Zimbabwe, Rwanda, Zambia, and Mozambique have all at times sent peacekeeping forces to try to dampen the conflict and prevent it from spreading, but with limited success.

Historically, the repeated recourse to war and armed conflict is evidence of the limitations of attempts to manage differences in society in a civilised fashion. Although at times states have built military aggression into their arsenal of foreign and domestic policies, for the most part societies try to avoid the breakdown of war. Throughout the twentieth century somewhat faltering attempts were made to develop international agencies that could act as a buffer and mediator between warring parties (see Chapter 12). However, their very partial success underlines the fact that maintaining a reasonable balance in the weight of military power between states and sub-state groups is still, ironically, usually a prerequisite for stability and peaceful coexistence. The most notable example of such a stand-off in the recent past is the Iron Curtain, that divided Europe for nearly half a century during the Cold War.

Key themes and further reading

The concept of difference is important for understanding the inherent instability built into states and all other governmental structures. Recognising difference and allowing for it within government are, therefore, crucial. The cohesive force of language needs to be appreciated, as does the dangers of racism and discrimination if ethnicity and cultural differences are not accommodated. The ways in which religious and political power are conflated within states are frequently fundamental to understanding how they function. The nature and meaning of nationalism, as well as the distinctions between state, sub-state, anti-colonialist, and ethnic and religious nationalism should now be clear. Ignoring these cleavages in society can encourage irredentism and lead to conflict and war.

Two very good introductions to the impact of cultural, racial, and linguistic differences on the dynamics of societies are: *Maps and Meaning* by Peter Jackson (1989) and *Cultural Geography* by Michael Crang (1998). Both can be recommended without reservation, though *Cultural Geography* has the advantage of being more up to date. To find out how peoples go about reconciling their differences and forging themselves into functioning political units, Norman Davies (1999) in *The Isles* gives a fascinating and very detailed interpretation of how the British Isles developed politically to be the way they are today.

6 Civil society, pressure groups, and political parties

> There is a consciousness on the minds of leading politicians that the pressure from behind, forcing upon them great measures, drives them almost quicker than they can go, so that it becomes a necessity with them to resist rather than to aid the pressure which will certainly be at last effective by its own strength.
>
> (Anthony Trollope, *Phineas Redux*, 1874, chapter 4)

Pressure groups and civil society

Civil society is a fundamental part of the fabric of the modern state, yet it is a difficult concept to pin down. What Urry (1981) has graphically described as the sphere of struggle is an essential interface between people and government, which allows new ideas to be brought to public attention, debated, and put on the political agenda. In a sense, it is a wider manifestation of the governance, as opposed to government, discussed in Chapter 4.

Civil society is usually thought of as achieving its most developed form in democratic capitalist societies, broadly Western democracies. Totalitarian societies of any political persuasion seek to smother dissent, rather than encouraging it as part of the democratic process, though while this is undoubtedly true in practical terms, it is not to say that covert dissent and, by implication, a covert civil society does not exist under dictatorships. When Saddam Hussein won 100 per cent of the popular vote in the presidential election in Iraq in October 2002, it clearly did not reflect the real mood of all the Iraqi people, but fear kept any opposition totally beneath the surface. The subsequent US-led invasion in 2003 and the multiplicity of social, religious, and political factions that have emerged in its wake are ample testament to the variety of different views in Iraqi society about how the country should be governed.

Environmentalism

The steady growth of environmentalism from its radical beginnings in the middle of the nineteenth century in the USA is a classic example of the way in which an alternative view of the world can gradually over time be incorporated into the mainstream political agenda. The most important initial spur was probably the publication in the USA of *Man and Nature* by George Perkins Marsh in 1864. It argued forcefully that there ought to be a much more sympathetic relationship between economic development and the environment, and that care for the environment should always be a key element of policy. The book caught the public imagination and certainly influenced campaigners like John Muir, who fought tirelessly to protect the environment of California, especially the magnificent redwood and giant sequoia forests, then as now under serious threat from commercial forestry (Figure 6.1). The crusade was also taken up in government by the USA's first professionally trained forester, Gifford Pinchot, who became Chief of the Division (now the Bureau) of Forestry at the US Department of Agriculture. He argued strongly for resource policies that ensured conservation and renewal and were sustainable in the long term. He was one of the most influential voices in rewriting the environmental agenda in the twentieth century (Runte, 1979).

Figure 6.1 ***Giant sequoias, King's Canyon National Park, USA***

Source: Robin Hill, with permission.

Similar ideas were also taken up in Europe, especially in interwar England through organisations like the National Trust, the Council for the Preservation of Rural England (now renamed the Campaign to Protect Rural England), and the national parks movement (Sheail, 1981; Lowe and Goyder, 1983). Very rapidly the values of environmental conservation began to spread worldwide and became a global cause, though not necessarily one embraced with any enthusiasm by governments.

During the middle part of the twentieth century environmentalism languished somewhat as a result of the Second World War and its aftermath, but in the 1960s there was a strong revival of interest. A number of key events occurred, most notably the publication of two books, *Silent Spring* by Rachel Carson (1962) and *The Population Bomb* by Paul Ehrlich (1968), which together produced a huge shock wave and alerted people to the serious and present dangers posed by pollution and overpopulation respectively. As a direct result of these and a number of other publications, the first UN Earth Summit was convened in Stockholm in 1970, putting environmentalism very much back on the political agenda and generating a host of organisations whose main purpose was to promote it as a political cause.

These same concerns also led to movements across many parts of the developed world calling for greater environmental justice (Cutter, 1995). At an international level, there is increasing concern and outrage at the way in which the burdens of pollution are being borne disproportionately by the less developed world and there are growing calls for the balance to be redressed (see also 'Transnational corporations' in Chapter 11). Nor is the disquiet restricted to the international level, within states too there is a growing body of evidence that it is the poor and politically less powerful that are routinely exposed to the risks of environmental hazards and pollution (Bullard, 1993). Particularly in the USA, there is a growing concern at the way in which those who are politically weak, which often means people of colour and immigrants, are exploited when it comes to such dangers, leading to charges of environmental racism.

The development of environmentalism as a coherent concept has been charted in detail by Tim O'Riordan in a series of books and articles since the 1980s (1981, 1989, 1996). In his view, environmentalism has evolved along two quite distinct lines (Figure 6.2). On one side are the *ecocentrics*, who are convinced that existing government structures are incapable of being adapted to meet the challenge posed by environmental degradation and change and that, therefore, there will have to be a fundamental redistribution of power towards a far more decentralised and small-scale economy, with much greater local public involvement and control. On the other side are the *technocentrics*, who believe that

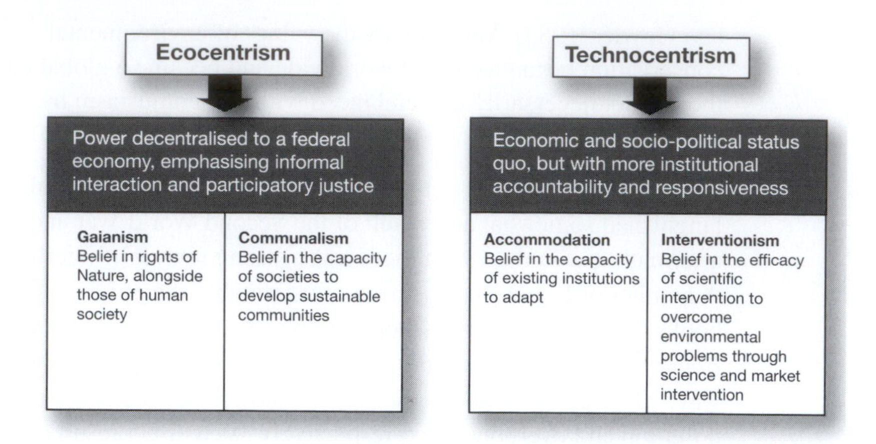

Figure 6.2 ***Forms of environmentalism***

Source: after T. O'Riordan (1989) 'The challenge for environmentalism', in R. Peet and N. Thrift, *New Models in Geography*. Unwin Hyman, London, pp. 77–101. Reproduced with permission of Thomson.

existing political structures can be modified to meet the needs of environmental conservation through technological solutions allied to improved regulation and education.

The incorporation of environmentalism into the political mainstream was given a huge boost by a second Earth Summit, the 1992 UN Conference on Environment and Development held in Rio de Janeiro in Brazil. This global event committed 118 countries to the twenty-seven principles of the Rio Declaration, thus ensuring that the world economy was managed in such a way as to promote sustainable development and the wise use of resources (Box 6.1). The key mechanism has been Agenda 21, a commitment to decentralised decision-making and the formulation and implementation of policies for sustainable development at all levels of government, from the world community, to the state, right down to the smallest division of local government. The true impact of Agenda 21 on policy-making is difficult to gauge, but it has certainly awakened people to the concept of sustainability, making the principles widely talked about, if not always acted upon. In Central and Eastern Europe, for example, as democracy was beginning to take root again in the early 1990s, Green parties and policies for sustainable development were initially very prominent. No doubt this was heavily influenced by the chronic environmental problems inherited from the Communist era, as well as Agenda 21, though their importance has subsequently faded somewhat in the face of apparently more immediate social and economic challenges (Tickle and Welsh, 1998).

Box 6.1

Rio Declaration on Environment and Development

The UN Conference on Environment and Development met at Rio de Janeiro from 3 to 14 June 1992. It reaffirmed the Declaration of the UN Conference on the Human Environment, adopted at Stockholm on 16 June 1972, and seeks to build upon it, with the goal of establishing a new and equitable partnership through the creation of new levels of cooperation among States, key sectors of societies and people. The conference will work towards international agreements which respect the interests of all and protect the integrity of the global environmental and developmental system. Recognising the integral and interdependent nature of the Earth, our home, the conference proclaims that:

Principle 1

Human beings are at the centre of concerns for sustainable development. They are entitled to a healthy and productive life in harmony with nature.

Principle 2

States have, in accordance with the Charter of the United Nations and the principles of international law, the sovereign right to exploit their own resources pursuant on their own environmental and developmental policies, and the responsibility to ensure that activities within their jurisdiction or control do not cause damage to the environment of other States or of areas beyond the limits of national jurisdiction.

Principle 3

The right to development must be fulfilled so as to equitably meet developmental and environmental needs of present and future generations.

Principle 4

In order to achieve sustainable development, environmental protection shall constitute an integral part of the development process and cannot be considered in isolation from it.

Principle 5

All States and all people shall cooperate in the essential task of eradicating poverty as an indispensable requirement for sustainable development, in order to decrease the disparities in standards of living and better meet the needs of the majority of the people of the world.

Principle 6

The special situation and needs of developing countries, particularly the least developed and those most environmentally vulnerable, shall be given special priority. International actions in the field of environment and development should also address the interests and needs of all countries.

Principle 7

States shall cooperate in a spirit of global partnership to conserve, protect and restore the health and integrity of the Earth's ecosystem. In view of the different contributions to global environmental degradation, States have common but differentiated responsibilities. The developed countries acknowledge the responsibility that they bear in the international pursuit to sustainable development in view of the pressures their societies place on the global environment and of the technologies and financial resources they command.

Principle 8

To achieve sustainable development and a higher quality of life for all people, States should reduce and eliminate unsustainable patterns of production and consumption and promote appropriate demographic policies.

Principle 9

States should cooperate to strengthen endogenous capacity-building for sustainable development by improving scientific understanding through exchanges of scientific and technological knowledge, and by enhancing the development, adaptation, diffusion and transfer of technologies, including new and innovative technologies.

Principle 10

Environmental issues are best handled with participation of all concerned citizens, at the relevant level. At the national level, each individual shall have appropriate access to information concerning the environment that is held by public authorities, including information on hazardous materials and activities in their communities, and the opportunity to participate in decision-making processes. States shall facilitate and encourage public awareness and participation by making information widely available. Effective access to judicial and administrative proceedings, including redress and remedy, shall be provided.

Principle 11

States shall enact effective environmental legislation. Environmental standards, management objectives and priorities should reflect the environment and development context in which they apply. Standards applied by some countries may be inappropriate and of unwarranted economic and social cost to other countries, in particular developing countries.

Principle 12

States should cooperate to promote a supportive and open international economic system that would lead to economic growth and sustainable development in all countries, to better address the problems of environmental degradation. Trade policy measures for environmental purposes should not constitute a means of arbitrary or unjustifiable discrimination or a disguised restriction on international trade. Unilateral actions to deal with environmental challenges outside the jurisdiction of the importing country should be avoided. Environmental measures addressing transboundary or global environmental problems should, as far as possible, be based on international consensus.

Principle 13

States shall develop national law regarding liability and compensation for the victims of pollution and other environmental damage. States shall also cooperate in an expeditious and more determined manner to develop further international law regarding liability and compensation for adverse effects of environmental damage caused by activities within their jurisdiction or control to areas beyond their jurisdiction.

Principle 14

States should effectively cooperate to discourage or prevent the relocation and transfer to other States of any activities and substances that cause severe environmental degradation or are found to be harmful to human health.

Principle 15

In order to protect the environment, the precautionary approach shall be widely applied by States according to their capabilities. Where there are threats of serious or irreversible damage, lack of full scientific certainty shall not be used as a reason for postponing cost-effective measures to prevent environmental degradation.

Principle 16

National authorities should endeavour to promote the internalisation of environmental costs and the use of economic instruments, taking into account the approach that the polluter should, in principle, bear the cost of pollution, with due regard to the public interest and without distorting international trade and investment.

Principle 17

Environmental impact assessment, as a national instrument, shall be undertaken for proposed activities that are likely to have a significant adverse impact on the environment of those States. Every effort shall be made by the international community to help States so affected.

Principle 18

States shall immediately notify other States of any natural disasters or other emergencies that are likely to produce sudden harmful effects on the environment of those States. Every effort shall be made by the international community to help States so afflicted.

Principle 19

States shall provide prior and timely notification and relevant information to potentially affected States on activities that may have a significant adverse transboundary environmental effect and shall consult with those States at an early stage and in good faith.

Principle 20

Women have a vital role in environmental management and development. Their full participation is therefore essential to achieve sustainable development.

Principle 21

The creativity, ideals and courage of the youth of the world should be mobilised to forge a global partnership in order to achieve sustainable development and ensure a better future for all.

Principle 22

Indigenous people and their communities and other local communities have a vital role in environmental management and development because of their knowledge and traditional practices. States should recognise and duly support their identity, culture and interests and enable their effective participation in the achievement of sustainable development.

Principle 23

The environment and natural resources of people under oppression, domination and occupation shall be protected.

Principle 24

Warfare is inherently destructive of sustainable development. States shall therefore respect international law providing protection for the environment in times of armed conflict and cooperate in its further development, as necessary.

Principle 25

Peace, development and environmental protection are interdependent and indivisible.

Principle 26

States shall resolve all their environmental disputes peacefully and by appropriate means in accordance with the Charter of the United Nations.

Principle 27

States and people shall cooperate in good faith and in a spirit of partnership in the fulfilment of the principles embodied in the Declaration and in the further development of international law in the field of sustainable development.

Concerns about the future of the environment have become embedded into the formal political process through a number of different channels. NGOs, such as Greenpeace and the Worldwide Fund for Nature, have been instrumental in promoting environmental causes internationally through campaigns on specific issues, such as that to stop whaling and to protect the habitats of other endangered species. Through their activities, they seek to influence national and international policy to come into line with their own agendas. There are also a multitude of environmental organisations focused on a more limited range of national issues, such as the Sierra Club in California, which has worked tirelessly for more than a century to protect the Rocky Mountains from commercial exploitation and to return the whole cordillera region to a state of pristine wilderness.

The activities of these, and a host of other organisations, have forced environmentalism into a position of much greater political prominence, which in turn has seen the emergence of Green parties in a number of countries, in some cases enjoying sufficient popular political support to enable them to participate in government at national level. In Germany, the Green Party has been in coalition with the Social Democrats since 1998, occupying several senior posts in the government, including the foreign office, and putting it in a position to have a major influence on the direction of international negotiations on the future of the environment.

High levels of popular support also have an effect on the behaviour of the public. Consumers demand goods and services that are environmentally friendly, which in turn encourages producers to demonstrate that their production processes do not offend environmental ethical principles. The true robustness of this commitment on the part of both consumers and producers is often open to question, faced with the competition of a global marketplace (see Chapter 11), but undoubtedly the public in most of the developed world expects there to be a much deeper and more explicit political commitment to environmental principles.

Along with the growing global concern for the environment, there has been mounting pressure for more real progress on environmental justice (Cutter, 1995). There has been a growing realisation that the burden of the consequences of environmental degradation has not been shared equally, falling disproportionately on the poor and marginalised, especially in the developing world (see 'Transnational corporations' in Chapter 11). They are at much greater risk of being exposed to the toxic effects of pollution and lack the resources to protect themselves. This has created a growing movement for environmental justice, especially in the USA. It is geared to exposing and combating the inequities to which people of colour, immigrants, and other groups who for one reason or another lack political clout, routinely have to suffer. Many environmental justice campaigns have also been concerned to bring to light what they see as the self-serving agenda of parts of the environmental movement, criticising it for seeking to limit access to resources for the privileged few, rather than for the many. This criticism has been particularly levelled at organisations whose main aim is landscape protection, because it is argued that the benefits are only available to very restricted portions of society.

Environmentalism is a particularly clear example of the way in which a rather esoteric minority philosophy can come to influence profoundly public attitudes and, eventually, government policy. Once a certain level of momentum is achieved in the ranks of civil society, governments have to take notice and either incorporate the new philosophy or counter it by argument or suppression. No matter how they respond, however, the process is invariably ongoing. Environmentalism, for example, has now begun to turn from wise resource use to other forms of human–environment relations. The rights of animals and a code of ethics related to the use and abuse of wild and domestic animals is a growing issue in the early part of the twenty-first century and, in many developed countries such as the UK and the USA, animal liberationists are a growing political force (Francione, 1996).

Voluntary bodies and social infrastructure

States and the infrastructure of government are inevitably the dominant influences in society, but they are self-evidently far from the only ones. Most societies, especially those with any pretensions to being viewed as democratic, are also at pains to encourage the church and the law as independent, counterbalancing institutions. Part of their role is to temper and keep a check on the power of the state, but in neither case is their prime function to cover those parts of the social infrastructure that are poorly served by both the public and private sectors. These gaps tend to be filled, at least partially, by a whole range of voluntary bodies, most of which emerge spontaneously and carve out niches for themselves, often rapidly becoming an indispensable part of the fabric of society. In many instances the value of voluntary bodies is also quickly recognised by government, which moves to support them by formally recognising their role and contribution by direct or indirect funding, or by other means.

In England and Wales the vast majority of such voluntary bodies are designated as charities under the control of the Charity Commission, which ensures that they only work in support of their designated charitable purposes, that they keep properly audited accounts, and that they do not engage in any activities that could be construed as political campaigning (Charity Commission, 2002). In return for operating within these prescribed limits, the government allows the organisations to reclaim any tax paid on their income, a powerful incentive for ensuring compliance. There are over a million registered charities in England and Wales, covering every conceivable type of non-political social activity and encompassing organisations from small local trusts with an annual income of just a few hundred pounds, to national bodies like the National Trust with over three million members, and the Royal Society for the Protection of Birds with over one million members.

Voluntary bodies are, thus, an integral part of the fabric of society, but they suffer from a chronic inherent weakness in that when it comes to providing a service their geographical range may be very restricted. This can be the result of their mission, which may be limited to a particular area or group, or to the funds available to them. Usually, it is caused by a combination of these and other factors, but the outcome is the same, namely uneven provision. Overcoming this limitation is usually very difficult and makes it hard for voluntary organisations to provide a geographically comprehensive level of provision and quality of service.

Citizens Advice

No organisation illustrates better the role that the voluntary sector plays in society than Citizens Advice, the operating name for the National Association of Citizens Advice Bureaux in England, Wales, and Northern Ireland (Blacksell *et al.*, 1991). As the umbrella body for over 2,000 independent bureaux, with 2,800 outlets, across the three countries, it coordinates a national network of free advice centres, largely staffed by volunteers and open to all on a first come, first served basis. Coverage across the whole of the UK is complemented by Scotland's equivalent organisation, Citizens Advice Scotland, which has 70 bureaux and 199 outlets (Figure 6.3).

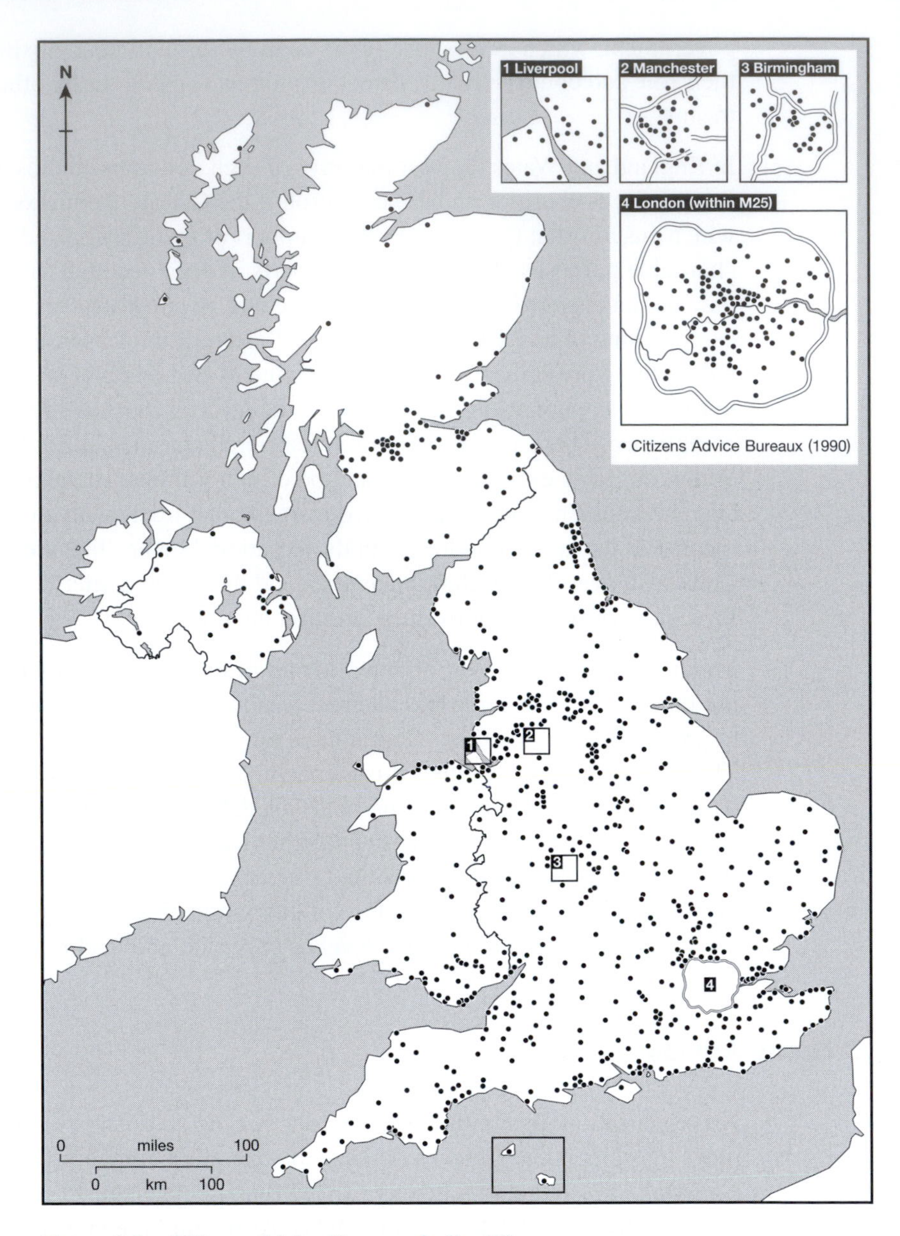

Figure 6.3 ***Citizens Advice Bureaux in the UK***

Source: M. Blacksell, K. Economides, and C. Watkins (1991) *Justice Outside the City*. Longman, London, p. 47. Reproduced with permission of Pearson Education.

The service has earned widespread respect for its professionalism and the quality of its advice services, and as long ago as 1979 the Royal Commission on Legal Services recommended that it should form the basis of the primary level of advice-giving nationally. Citizens Advice Bureaux (CABx) have since become an accepted part of the national infrastructure of social provision in the UK and, in 2002–3, they dealt with more than 6 million client problems covering a wide range of social and economic issues. The main focus has always been on helping clients navigate through the labyrinth of social security and housing benefit systems, ensuring that they receive their full entitlements, but other areas, such as employment law and managing personal debt, have also become increasingly important. The quality of the advice offered is recognised to be of such a high order that invariably the intervention of a CAB adviser brings immediate resolution of a dispute. Indeed, it is not at all uncommon for central and local government agencies to actually refer to the CAB service to help resolve difficult cases. It is no exaggeration to say that the service is a prime example of the voluntary sector providing an indispensable initial gateway to a raft of statutory services in the UK.

The scale of CAB operations and the value placed upon them is immediately apparent from their level of public funding. In 2002–3 their total budget in the UK was in excess of £150 million, more than 50 per cent of it coming from local authorities, with most of the rest coming from central government and EU sources. Although the service guards jealously its independence and its predominantly voluntary ethos – in 2002–3 there were more than 20,000 volunteers as opposed to just over 5,000 paid staff – such a heavy reliance on government funding makes it hard for it not to be seen in practice as part of the public sector social provision.

The voluntary ethos at the heart of the original concept is, therefore, proving increasingly difficult to sustain as the expectation rises that standards will be of the highest professional order and the level of public funding continues to increase (Blacksell *et al.*, 1990). All volunteers now have to undergo rigorous training and the standard of advice provided is subjected to the sharpest scrutiny by paid professional staff. Finding volunteers of sufficient calibre, who are prepared to assume such a high level of responsibility and commitment and who live in the places where they are needed is extremely challenging. Inevitably, the areas that tend to lose out are the more sparsely populated and remote rural areas, and other places where access poses problems (see Figure 6.4). It has even been seriously debated whether there should not be paid volunteers, a

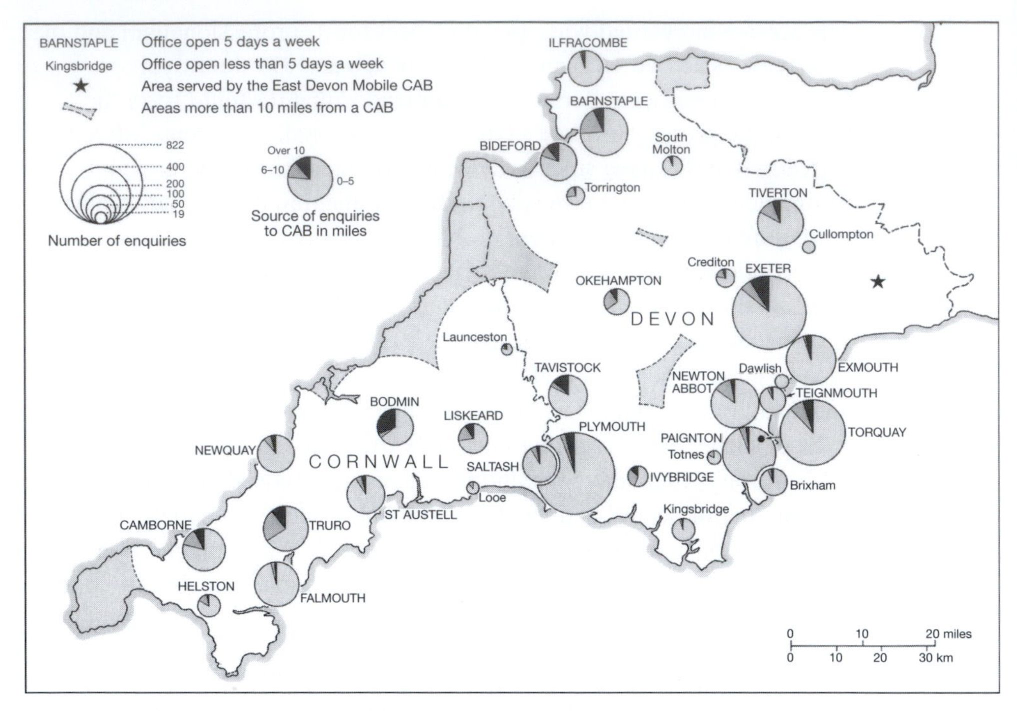

Figure 6.4 ***Demand for CAB services in Cornwall and Devon***

Source: M. Blacksell, K. Economides, and C. Watkins (1991) *Justice Outside the City*. Longman, London, p. 109. Reproduced with permission of Pearson Education.

rather contradictory concept, but clear evidence of the crisis facing the CABx and other voluntary services (Blacksell and Phillips, 1994).

Nevertheless, in his review of 2002–3, David Harker, the Chief Executive of Citizens Advice, proclaimed confidently that: 'A strong civic society needs a CAB' (Citizens Advice, 2003) and the service illustrates vividly the key contribution that the voluntary sector makes to the governance of society. The difficulties facing the CAB as a service dependent on volunteers also reveal the dangers of relying on it too much: if the volunteers are not forthcoming in a particular area, for whatever reason, then the service cannot be provided; the struggle for funding always threatens the scope of the service; and offering the service completely free, although it enhances access in one respect, is also a restriction, because it means that those using the service have no direct control over whether or not it is available.

Political parties

Political parties are the ultimate pressure groups, seeking to become the government, or at the very least have a share in it. They are the oil that enables the political process to function smoothly. Although public representation and regular elections are essential for democracy, they are rendered largely meaningless in the absence of organised political parties. Indeed they have been described as 'the primary institutions of representative democracy' (Dalton, 1988, p. 127).

People view political parties as the main agency through which they can influence government. At election time, parties are central to the whole process and governments rely on their support for retaining power. In societies new to democracy, such as those in Central and Eastern Europe in the early 1990s after the collapse of Communism, the creation or re-establishment of a range of different parties to replace the tyranny of the one-party state was one of the most pressing tasks in the process of social reconstruction (Klingemann and Fuchs, 1995).

Political parties must, therefore, reflect all the different views in society, while at the same time simplifying and consolidating them, so that a degree of order can emerge. They have, therefore, been at the forefront of the struggle in Europe to achieve evolutionary rather than revolutionary change. When political parties have been weak, as in the short-lived and ill-fated Weimar Republic in Germany between 1919 and 1933, or in the former Yugoslavia after the death of Josip Broz (Marshall Tito) in 1980, totalitarianism has tended to have the upper hand. When they have been strong, as in most of Western Europe since the middle of the twentieth century, they have generally been a creative and positive force, underpinning political stability (Blacksell, 1998).

Most political parties never succeed in forming a government, though all probably dream of the possibility, and see their main role as representing different shades of opinion to government. Historically, many have little success in doing even this, acting as no more than a safety valve for specific groups and interests, though as societies have become ever better educated and sophisticated, minority parties have developed increasingly effective means of making their voices heard and ensuring that their agendas are incorporated into government policy.

The influence of political parties is by no means always viewed by governments as benign and action to ban them is a regular, if infrequent, occurrence. The right-wing Socialist Reich Party (SRP) and the

Communist Party (KPD) have been banned in the Federal Republic of Germany since 1952 and 1956 respectively, because they are seen as being hostile to the *Grundgesetz*, the German Constitution. In Northern Ireland, Sinn Féin, the nationalist party that campaigns for a unified and independent Ireland, has been banned in the past for its advocacy of violence and the force of arms to achieve its political ends. Similar restrictions have been imposed by the Spanish government on ETA, the military wing of the Basque nationalist party, Herri Batasuna, and by the French government on the FLNC, the Corsican independence party, and there are many other examples.

Outlawing political parties is, however, no panacea. If a banned party actually commands substantial public support, it can become a vehicle for revolution and the overthrow of the established political order. The 1917 Revolution in Russia saw the Communist Party move directly from illegality to power and the consequences of that lesson have not been lost since on governments elsewhere in the world. They ignore the force of public opinion at their peril and, where outlawed political parties are steadily gaining support, ways and means are almost always found to modify policy and incorporate them into the political mainstream, though sometimes with disastrous results, as in Germany with the Nazi Party between the two world wars.

Regionalism and political parties

The most striking feature of the way in which political parties are organised is the extent to which they are tied to national boundaries. Every country has its own collection of parties, the majority of which have at best very weak links beyond its frontiers. Equally, the realities of power mean that unless parties aspire to achieving a national mandate they will be consigned either to opposition, or to a relatively minor supporting role in government. Nevertheless, regionalism in party politics is important for two reasons. The first is that party popularity is frequently uneven across a national jurisdiction and, although a party may not be part of the government, it is perceived as representing particular regional interests. The second is that parties are often formed specifically to represent and promote a regional minority.

The effects of uneven national support for political parties in Italy, especially after the electoral reform that preceded the 1994 elections, show that only the socialist PDS (*Partito Democratico della Sinistra*) and the newly formed *Forza Italia* were truly national parties (Agnew, 1996). All the other significant players depended heavily on regional power bases, tainting their credentials as guardians of the national interest. The most obviously regional of these minority parties is the Northern League (*Liga Nord*), which was formed in 1991 by amalgamating a number of small parties in the north of Italy, but the National Alliance (*Alleanza Nationale*) also has a clear regional focus, being heavily concentrated in the *Mezzogiorno* to the south of Rome. The astonishing success of *Forza Italia* in 1994, when it emerged from nowhere to seize power, was based largely on the way its leader, Silvio Berlusconi, managed to forge workable electoral alliances with both the Northern League and the National Alliance, thus bolstering support in those areas where it was weakest (Agnew, 1997). Nor was this success just a flash in the pan; despite losing popularity and power in the late 1990s, it won back power and again formed a national government in 2001.

It is always difficult to encompass the full gamut of regional minorities, even those that are sufficiently organised to have a political party to represent their interests. The main stumbling block is the enormous variation in the internal organisation of central states themselves. In Europe, there is at one extreme the case of Switzerland, where decentralisation to *canton* level is such that the national government has only limited power and the main rivalries are between the *cantons* themselves. At the other, in the UK for example, where there is heavy emphasis on central decision-making, regional dissatisfaction is very much more pronounced.

Notwithstanding these difficulties, Figure 6.5 attempts to depict those areas where regional parties have gained significant political power in Europe. In a few cases they have managed to forge an alliance with a national party and have created a genuine national power base, the outstanding example of such manoeuvring being the alliance between the Christian Socialist Union (CSU) in Bavaria and the national CDU. This has endured for more than half a century and the only other alliance approaching this success is the Northern League in Italy, though it is nowhere near the secure position achieved by the CDU/CSU coalition in Germany.

More commonly, regional parties prefer to act more as pressure groups for change. In Belgium, a country chronically divided by both religion and language, both the Flemish majority in the north and the French-speaking Walloon minority in the south have their own nationalist

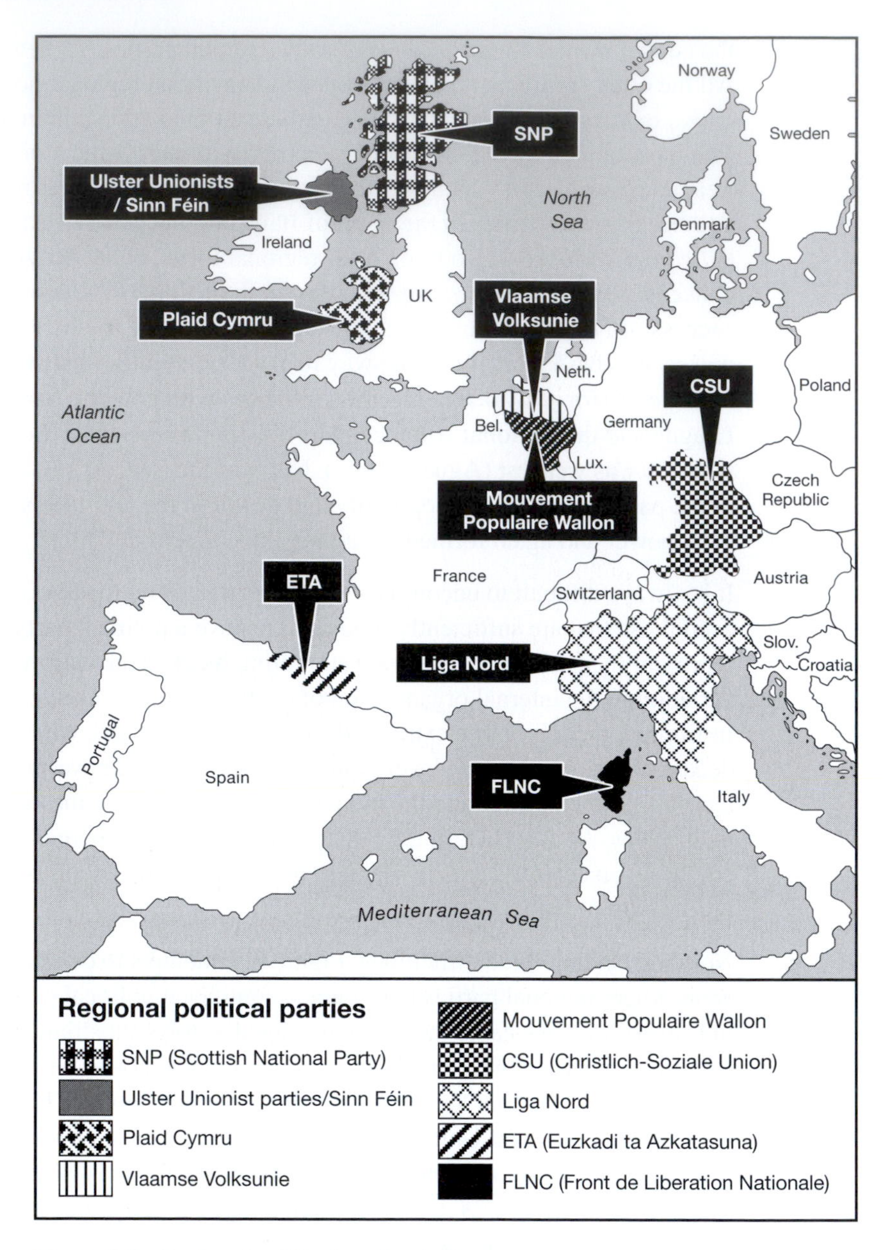

Figure 6.5 ***Regional political parties in Europe***

Source: T. Unwin (1998) *A European Geography*. Longman, Harlow, p. 123. Reproduced with permission of Pearson Education.

parties. In the early years of the twentieth century, Flemish separatism was much more strident in the face of French domination of the education system and the civil service, but since the 1950s they have used their majority steadily to turn the tables. In terms of both the economy and society, the Flemish north is very much in the ascendancy and its regional party, *Vlaamse Volksunie*, is in relative abeyance. By contrast, the Walloons in the south are represented by several regional parties, all of which have, or have had, elected members in the national parliament in Brussels. The most important is the *Mouvement Populaire Wallon*, founded in 1961, but has subsequently been joined by the *Partie Wallon des Travailleurs*, the *Front Commun Wallon*, the *Front Démocratique des Francophones*, and the *Rénovation Wallonne* (Lorwin, 1966).

The European Parliament

Despite the dynamism of economic integration in Europe, not to mention cooperation in defence matters and social affairs, it is surprising that party politics have remained somewhat uncompromisingly national. Nevertheless, there are broader trends and slowly they are becoming a more significant element in the political landscape. The touchstone for this has been the European Parliament, which for all its historical constitutional weakness is gradually growing in importance (Bardi, 1996). Partly this is simply a reflection of the growth in membership of the EU itself. The original six signatories had a combined population of over 190 million, but with twenty-five countries it now embraces 457 million people. It also, however, reflects a change of status. Initially the members of the forerunner of the European Parliament were appointed by member states, but since 1979 six successive direct elections, the most recent in 2004, have considerably enhanced its stature, even though its powers are still very limited in comparison with those of national parliaments and there are still widespread complaints of a 'democratic deficit' at the heart of the EU (Lodge, 1991 and 1996) (Box 6.2).

The most recent elections for the European Parliament were held in 2004, returning 732 MEPs from the twenty-five member states, the number from each depending on the size of its population (Figure 6.6). The parliament incorporates 171 different political parties and for it to be a workable institution, it has had to align itself into larger, simplified groupings. Seven major clusters have emerged, the largest being the

Box 6.2

The European Parliament, 2004

Since the elections in June 2004, the European Parliament has had 732 Members elected in the twenty-five states in the EU for a five-year term of office. Most of the time, the Parliament and the MEPs are based in Brussels where its specialist committees meet to scrutinise proposals for new EU laws. For one week a month the Parliament meets in Strasbourg in full plenary session to amend and vote on draft legislation and policy. The headquarters of the Parliament's civil service is located in Luxembourg, though many of its officials are based in Brussels.

In addition to their growing role as legislators, MEPs approve the appointment of the European Commission, decide the EU budget with member states, monitor spending, approve international agreements, question EU Commissioners, and appoint the European Ombudsman. Citizens have the right to petition the European Parliament.

MEPs do not sit in national delegations in the Parliament, but in multinational political groups. The centre-right European People's Party (EPP) and the European Democrats, which includes the British Conservatives, is the largest political group. British Labour MEPs belong to the Party of European Socialists, the second biggest group. The Alliance of Liberals and Democrats for Europe (ALDE), where the largest national contingent is from the UK, is the third biggest, closely followed by the Green/European Free Alliance which has brought together Green MEPs from 13 countries, including Britain's two Green members, and nationalist parties, including Plaid Cymru and the SNP.

British members elected for the UK Independence Party (UKIP) joined the Independence and Democracy Group. The other groups are the European United Left/Nordic Green Left, which brings together the Left and Green members from a number of countries, including France, Greece, and Italy; and the Union for Europe of the Nations. There is also a small non-aligned group whose members sit as independents.

EPP–ED Group (European People's Party – the Christian Democrats – and the European Democrats) with 268 members on the political right, and the PES (European socialist parties) with 200 members on the left. In the centre sits the ALDE (Alliance of Liberals and Democrats for Europe) with 88 members. There are then smaller groups on both the right and left wings of the political spectrum, including the Green–European Free Alliance and the Independent Democrats, both of which are left of centre, as well as 29 completely non-aligned members.

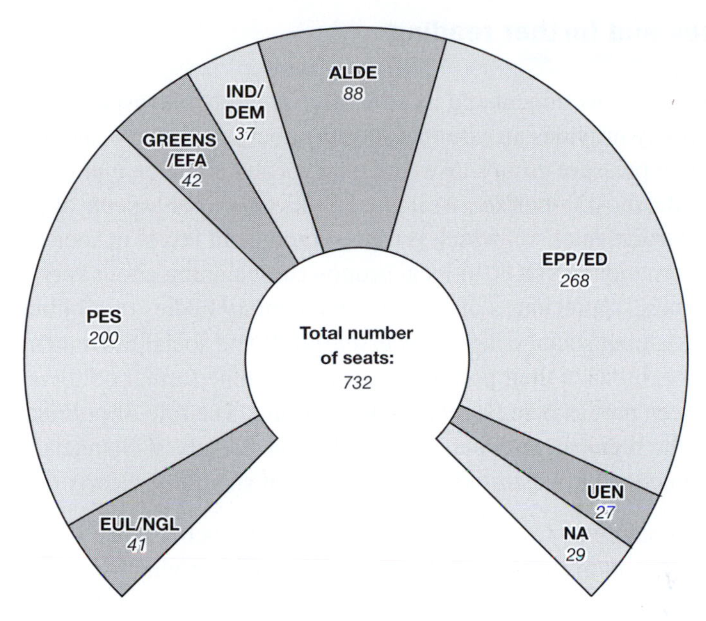

EUL/NGL - Alliance of seventeen left-wing parties
PES - European socialist parties
Greens/EFA - Green/European Free Alliance
IND/DEM - Independent Democrats
ALDE - Alliance of Liberals and Democrats for Europe
EPP/ED - European People's Party (Christian Democrats)/European Democrats
UEN - Union for Europe of the Nations
NA - Non-aligned

Figure 6.6 ***Membership of the European Parliament, 2004***

The European Parliament, in common with all those in other parts of the world where there is a large number of parties represented and a governing majority has to be fashioned through negotiation and bargaining, only gels as a governing force after the elections themselves are over. The various national parties all campaign on national issues during the elections with little reference to the eventual structure of the parliament itself. Producing an effective democratic force out of such a complex mix of interests is thus extremely difficult, though if the experience of most national parliaments is taken as a guide, then there will be a process of gradual amalgamation over time and the tentative party groupings will steadily become more formalised. If this does not happen, then the parties are poorly placed to exercise power and there is a danger that their national electorates might become disillusioned and either desert them for one of their better organised competitors, or search for other, perhaps violent, ways of making their voices heard.

Key themes and further reading

Formal government and its structures are only one aspect of states; civil society provides an essential interface between people and government. Such pressure groups take many forms and embrace many causes. One of the most important, with close links to several aspects of geography, is environmentalism, which is represented at all levels in society, from international NGOs, to local groups campaigning about very specific issues. Appreciation of the role of voluntary bodies of all kinds in sustaining social cohesion is essential. Some social movements, over time, broaden their political base and become formal political parties, Green parties being a recent case in point. The role of political parties at the regional, national, and international levels as channels for reflecting public attitudes in government should be clearly understood.

The nature and workings of civil society is such a broad topic that there are any number of good, but very different, introductions to it. However, a very clear and well-argued analysis is provided by John Urry (1981) in *The Anatomy of Capitalist Societies*. Environmentalism is likewise a broad church, but one to which geographers have contributed much more fully and directly. By far the most thorough and thought-provoking treatment of the subject is *Environmentalism* by Tim O'Riordan (1981). The role of law and voluntary organisations in civil societies is discussed by the author, together with Kim Economides and Charles Watkins in *Justice Outside the City* (1991).

7 Electoral geographies

> An election is coming. Universal peace is declared, and the foxes have a sincere interest in prolonging the lives of the poultry.
>
> (George Eliot, *Felix Holt*, 1866, chapter 5)

Voting, the widening franchise, and the growth of electoral geography

Elections and the way they are organised are fundamental to the make-up of all democratic societies; they also profoundly affect their social and political geography, in that they provide the means by which people are represented in government. Voting gives people a voice, but the way constituencies at all spatial scales – local, regional, national, and even supra national – are organised is critical in deciding how their voices will be heard and acted upon. As George Eliot noted in her novel *Felix Holt*, elections are the one real opportunity for the people to call to account the executive that normally rules their lives. There are, of course, many other factors which also influence elections, in particular who is eligible to vote, the voting system used, and the frequency with which elections are held, but also the size and shape of constituencies is critical, because quite small demographic shifts can impact radically on the distribution of political power.

It is, therefore, quite surprising how relatively recently electoral geography has established itself as a core element of political geography. Aside from a few isolated research monographs, it was only in the late 1970s that there was any systematic interest shown by geographers in the subject (Taylor and Johnston, 1979), but the upsurge of interest in the UK then sparked not just a great deal of interest in the spatial patterns produced by election results, but also theoretical research on the significance of electoral studies for understanding the underlying dynamics of political geography (Johnston, 1979; Gudgin and Taylor, 1979).

Certainly the lack of attention paid to voting patterns by geographers was not the result of a general lack of interest in the subject. Elections have been pored over in detail by political scientists ever since representative government began to become the norm in the Western world at the end of the eighteenth and beginning of the nineteenth centuries. In the UK this interest dates back to the Reform Act of 1832, which made the first serious attempt to ensure that the franchise in Britain more fairly reflected the distribution of population (Thrasher and Rallings, 2000) (Box 7.1). Not that the first Reform Act itself provided for anything approaching a universal franchise. In common with most democratic societies, the UK has inched towards this goal over a period of nearly two centuries, grudgingly conceding the right to vote to more and more of its people, group by group. Gender and ownership of property were initially the key qualifications. In common with most other countries, the franchise in the UK included only men who owned property, though this was then extended to leaseholders and, subsequently, to most working men. By the end of the nineteenth century, voting was more or less a right for all men over the age of 21, but women were completely disenfranchised. Not until 1918 did women gain the vote and, even then, only those over 30. It was ten years later that voting rights were finally extended to all adults over the age of 21. Since then, the only important change in the franchise has been the reduction in the age limit to 18 in 1970.

At the beginning of the twenty-first century, most Western democracies have a universal franchise and this general extension of the right to vote has made elections very much more geographically relevant. They are now more truly tests of public opinion as a whole, rather than measures of particular sectional interests, even though participation rates are sometimes surprisingly modest, throwing into question the extent to which they are true tests of the overall public mood. In Western Europe, voting is compulsory only in Belgium, Italy, and Luxembourg, though it was compulsory in the Netherlands until 1971. Not surprisingly, these countries have amongst the highest levels of voter participation, consistently running at over 80 per cent of the electorate, but there are other countries, such as Iceland and Sweden, where voting is not compulsory and which have similar levels of participation. The missing voters in those countries where voting is compulsory are explained by

Box 7.1

The Reform Acts

The three Reform Acts of 1832, 1867, and 1884 all extended voting rights to previously disenfranchised citizens. Most controversial was the 1832 Act, which reapportioned representation in Parliament in a way that was fairer to the rapidly expanding industrial cities of the north, and did away with the so-called 'rotten' and 'pocket' boroughs. Old Sarum near Salisbury, for instance, only had seven voters, all controlled by the local squire, yet still sent two members to Parliament. The act not only reapportioned representation in Parliament, so that it more accurately represented the distribution of population, it also extended the franchise to those who were less powerful, socially and economically. The act extended the right to vote to any man owning a household worth £10, thus adding 217,000 to an electorate of 435,000 and giving one man in five the right to vote.

The 1867 Act extended the right to vote still further down the class ladder, adding nearly a million voters, mostly working men, to the electorate and bringing the total number eligible to vote to two million in England and Wales. The 1884 Act, together with the 1885 Redistribution Act, further tripled the electorate and gave the vote to agricultural labourers, by which time voting had to all intents and purposes become a right to all adult men. Women had to wait until the twentieth century to gain similar rights.

spoiled ballot papers, as many as 20 per cent of the electorate choosing not to cast their vote for any of the candidates on offer. At the other end of the scale, the proportion of the electorate voting in national elections in some countries is often well below 70 per cent, the lowest vote in Europe being 45 per cent in Switzerland in 1991. Turnout is usually even lower in local elections with turnouts of below 30 per cent quite common, and in the elections to the European Parliament the figures have been similarly low.

From a geographical point of view, it also needs to be remembered that the voting system used can have a significant influence on the eventual outcome. That most commonly used is the straightforward one person one vote across a jigsaw of constituencies. Under this first-past-the-post system, those gaining the most votes are duly elected to serve the particular constituency in question. The limitation is that, if there are more than two candidates, then the persons elected will more than likely only have gained a minority of the votes cast and the wishes of the majority of voters will have been ignored. It also means that no politician can ignore the minutiae of their local constituency affairs, even if they are fully engaged on the national or international stage, and none can take winning their constituency seat for granted.

The single transferable vote is one way of ensuring that voter wishes more nearly match who is eventually elected. Under this system voters rank the candidates and then a process of elimination occurs. At each iteration, the candidate with the least votes is eliminated and the first-choice votes cast for them are redistributed to the voters' second choices. The process ends when one candidate has 50 per cent of the votes cast. It is undoubtedly a fairer system, although it too has its disadvantages, notably the fact that voters are encouraged to rank candidates, even though they may know very little about what some of them stand for. It is also a very laborious and slow process to count the votes, but those areas where the single transferable vote is used, such as Northern Ireland, it is a popular system and has succeeded in broadening the political range of candidates elected.

The potential conflict between national and local party loyalties has been tackled in a different way using a list system, whereby the number of candidates elected for a particular party is determined by the proportion of the votes cast for that party. In some instances, such as the Federal Republic of Germany, there are national lists covering the whole country, in others, such as the elections for the European Parliament, there are constituency lists. The problem with the list system is that there is not much direct affinity between the candidates and the voters, who often have little idea who it is who is representing them. The criticism is countered by the argument that there are invariably other, lower tiers of government using the first-past-the-post, or the single transferable vote systems and the candidates elected at these levels can adequately represent individual voters.

Whatever the limitations of elections as measures of public opinion, be it through low levels of participation, or the inherent unfairnesses of the voting systems used, they do provide unparalleled insights into the way societies are changing. People are increasingly moving much more frequently, both within and between countries, so that the socio-economic structure of voter groups may change quite dramatically from one election to the next. Demographic factors, such as birth and death rates and rates of household formation are also constantly changing, and they too will affect what people want from society and the way they vote. More sinisterly, the manipulation of elections, through where exactly boundaries are drawn and which particular segments of the population are included, or not included, within constituencies can materially influence who is elected.

The spatial organisation of elections

Ensuring that elections are fairly and properly managed is one of the most important public duties in a democratic society. All countries with representative forms of government try to ensure that they have in place unambiguous rules for the conduct of elections, as well as the means to enforce them. The task is extremely difficult, because bias is all too easy to hide amid constant social change, but the price of failure is high. Once particular groups or sections of the electorate have been favoured, it is extremely hard to persuade them to relinquish their position and a monopoly of political power and systematic discrimination against those excluded is the result. If not addressed, this can become the source of deep-seated and destabilising social unrest, in extreme cases threatening the stability of the state.

The electoral goal in most countries is to devise and implement a system where every adult of voting age has a single vote of equal weight to all others, following the principle of one person one vote. In addition, constituency boundaries must be drawn in such a way that each includes roughly the same number of people and that, as far as possible, everyone has a similar social mix, a procedure known as districting. Of all countries, the USA has probably legislated most vigorously, though not necessarily most successfully, to achieve proportionality. The number of people eligible to vote per electoral district is regularly reviewed after each census to ensure that none is over or under represented. The US Supreme Court has repeatedly ruled unconstitutional even very small deviations which meant that there was not absolute equality of voters in each district. Most states subscribe in principle to proportionality, but are prepared to tolerate a greater degree of limited deviation in the interests of effective representation. This was the reason given for creating additional constituencies for the 1991 elections in Zambia. In order to ensure that areas with severe communications difficulties were not ignored, a rural bias was engineered and they were allocated a voting strength out of proportion to their population. The process was open and accepted by all sides in the election, but it still meant that some votes were devalued. It goes without saying that the whole process of districting is dependent on the availability of accurate and up-to-date population statistics, something that even countries using the most sophisticated census techniques struggle to achieve (Westminster City Council, 2002). Indeed, in the USA reapportionment used to be undertaken after every Congressional election, but because of arguments about the accuracy of population statistics, it now occurs after the US decennial census.

In the UK the task of ensuring electoral fairness has been the responsibility of the Electoral Commission since 2002, under which there are separate Boundary Committees for England, Northern Ireland, Scotland, and Wales. These committees, previously known as the Local Government Commissions and answerable directly to central government, have extremely wide-ranging powers including a duty to conduct periodic reviews of all constituency boundaries to ensure that demographic changes and population movements do not result in bias and imbalance in the system (Rossiter *et al.*, 1999a). Details of the Electoral Commission's responsibilities are set out in Box 7.2.

Box 7.2

The Electoral Commission

The Electoral Commission is an independent body set up by the UK Parliament under the terms of the Political Parties, Elections and Referendums Act 2000. Its brief is to increase public confidence and participation in the democratic process within the UK, by modernising the electoral process, promoting public awareness of electoral matters, and regulating political parties.

On 1 April 2002, the Boundary Commission for England, which had formerly been known as the Local Government Commission for England, became a statutory committee of the Electoral Commission. Its main duty is to keep local electoral boundaries under periodic review, to ensure that they reflect as far as possible the distribution of population. There are separate Boundary Commissions for Scotland, Wales, and Northern Ireland. In Scotland, the Boundary Commission is responsible to the Scottish Parliament; in Wales to the Welsh Assembly; and in Northern Ireland to the Northern Ireland Office, though this will change if local government is once again devolved to the province.

Malapportionment, or violation of the norm of equal representation according to population, is an ever-present danger, which those in charge of managing elections must constantly watch for and take steps to eliminate. In the USA, as far back as 1842, the Reapportionment Act required that Congressional Districts be contiguous and compact to prevent perverse and discriminatory boundaries being drawn, thereby excluding, for example, certain concentrations of unwanted economic or ethnic groups. The Act was strengthened further by a Supreme Court ruling in 1962, which held that Districts must have fair borders and encompass an appropriate population mix. A further ruling by the Supreme Court in 1985 deemed that any attempt to manipulate District borders to give advantage to one particular political party was unconstitutional. Despite such actions, however, the risk of malapportionment, intentional or unintentional, remains and the onus should always be on those responsible for managing elections to show that this form of bias is minimised.

There are many ways in which constituency boundaries can be intentionally manipulated for electoral advantage. Such acts are generally known as gerrymandering, after the Governor of the US State of Massachusetts, Elbridge Gerry, who crafted a District that bore a strong resemblance to a salamander for the political advantage of his own party. The ruse was spotted by his opponents, who called it a gerrymander after the perpetrator, a term picked up and publicised by the press. Suddenly there was a graphic word to describe the widespread abuse of power through spatial fraud and gerrymandering rapidly became established in the English language across the world.

All gerrymandering ultimately has the goal of unfairly encompassing an undue proportion of voters from one political party and there are a host of ingenious ways of doing this. The most common and blatant is to draw boundaries in such a way that the opposition is concentrated into a small minority of constituencies, thus reducing, and even eliminating, their chance of ever gaining political power. Social policies and discrimination have been used in the past through ethnic and religious ghettos, a notorious example being the concentration of the Catholic minority in Northern Ireland after the partition of Ireland in 1921 and the subsequent civil war.

A second commonly used stratagem is precisely the reverse strategy, whereby the voting power of the opposition is diluted across as many constituencies as possible, thus preventing them from ever achieving the number of elected representatives that would reflect their true electoral strength. In the UK the Liberal Democratic Party and its predecessor the Liberal Party have both campaigned long and hard for some form of proportional representation through a candidate list system to overcome this acknowledged bias. Not surprisingly, the Labour and Conservative Parties, which have monopolised political power since the early part of the twentieth century, have paid little more than lip-service to the Liberal Democrat pleas for greater fairness. The only major concession has been the introduction of a candidate list system in the elections for the European Parliament in 2004.

Both the above techniques of gerrymandering are difficult to prove and hard to eliminate, unlike a third technique, sometimes referred to as the stacked method, whereby small areas that are physically separated are lumped together to form single constituencies favouring the party in power. Invariably, however, attempts to stack votes in this way are so obvious that they rarely succeed.

There have been many attempts to counteract gerrymandering by using computer algorithms to generate constituencies automatically, but success has been very limited. It is not just the shortcomings of the algorithms themselves that pose problems, but also the almost endless scope for challenge by any party that feels it has been disadvantaged. It is a salutary lesson to understand that unless people have confidence in such theoretically unbiased results, the outcome can be just as controversial as more malevolent attempts to draw maps of electoral boundaries.

Finally, reference must be made to the most widespread of all election malpractices, voter registration. Unless voters are registered, it is impossible to conduct fair elections and ensuring that people do register, or are not being prevented from registering, is extremely challenging. Even in the most advanced democratic societies, there is great variation in the proportion of people registered to vote across countries, and in less developed countries it is often even more difficult to ensure the all those eligible are included.

The Westminster City Council gerrymander

A notorious attempt to gerrymander was perpetrated by a group of senior Conservative Councillors, led by Dame Shirley Porter, on Westminster

City Council in central London during the 1980s. Their particular ploy was to use the newly introduced right for council tenants to purchase their houses or apartments to bring more people who were likely to vote Conservative into key marginal wards and thus make it more likely that the ruling party would retain power. The irony of the case was that initially the social policies of the Council had been much praised, especially its 1987 decision to sell 500 homes a year as part of a 'building stable communities' initiative. Only later, under persistent probing by the National Audit Office, did the more sinister political agenda become apparent. The case eventually went to court and Dame Shirley was personally convicted and fined £42 million in surcharges, interest, and costs. She appealed successfully against the conviction in 1999, but this was overturned again by the Law Lords in 2001 and the case was finally settled with a reduced payment of £12 million in 2004. The whole saga illustrates very clearly how insidious gerrymandering can be, and how difficult it is to prove and to bring those responsible to book.

Elections and social change

Elections are good barometers of social change, because they directly involve a greater proportion of the public than any other test of opinion. They expose underlying social cleavages and often can provide a clear indication of the ways in which states and regions, at all levels, are changing. Furthermore, the longer established and more mature the democratic process is, the better election data are for uncovering changes of political mood. It is argued that political parties normally go through a three-stage process of development: from cadre parties, largely reserved for wealthy male elites; to mass parties representing particular sectional interests, such as the working class, or the church, or small farmers; to the all-embracing stage, found widely in Western Europe and North America today, where they try to appeal to all sections of society, irrespective of sectional affiliation (Panebianco, 1988). Once political parties have reached this final stage and begun to achieve a universal appeal, election results become a very sensitive measure of shifts in the public mood.

In the USA, two major political parties, the Democrats and the Republicans, have exercised a monopoly of power for most of the past two hundred years and shifts in the balance of power between them across the country are widely seen as reliable indicators of social and political priorities. Archer and Taylor (1981) made a fascinating analysis of the four-yearly presidential elections over 150 years and identified from the electoral geography a number of key turning points, such as the election of Franklin D. Roosevelt at the height of the economic depression in 1933, following which the underlying cleavages in American society experienced a quantum change as a result of the New Deal programme he inspired (Box 7.3).

Box 7.3

The New Deal

The New Deal is the name for a raft of government policies developed by President Roosevelt immediately after his election as the US President in 1933 to try to lift the USA out of the Depression. The strategy was basically to initiate a huge programme of public works across the country. A quarter of a million unemployed young men were offered jobs in forestry and flood prevention work, nearly twice as many as were at that time in the US regular army. The Tennessee Valley Authority was set up to revitalise the whole of the river basin of the Tennessee River, one of the rural areas worst affected by the Depression. Legislation was passed by the Congress to ease the burden of debt for farmers and homeowners, thus preventing millions of foreclosures. A huge sum at the time, $500 billion, was granted by the Congress to help states, counties, and municipalities in their duty of care for those who need direct and immediate relief. Finally, the Congress passed legislation allowing those states that wished to the right to sell beer. This resulted in many new jobs being created and a useful new source of tax revenue for the Federal government.

The initiative proved both popular and effective and a second wave of legislation extended further the ability of the Federal government to undertake public works. The Farm Relief Bill enabled huge amounts of public money to be poured into agriculture; legislation was enacted to enable working conditions for industrial workers to be regulated; and the Railroad Bill introduced a degree of central planning into the national railway network.

The New Deal as a whole was viewed by many in the USA as creeping socialism and, for that reason, was treated with great suspicion, but its success in creating a genuine economic revival meant that such fears were never allowed to derail the programme. In any case, the US involvement in the Second World War after 1942 squashed any argument about the rights and wrongs of direct government participation in the productive economy, as manufacturing industry turned increasingly to producing war materials.

Shifts such as those identified by Archer and Taylor go beyond the local effects that are evident in some form in all electoral data. Local effects can arise from some kind of genuine regional solidarity, like that traditionally accorded to the Democratic Party in the American South in the south-eastern USA, but they can also be the result of cynical manipulation. Governments always run the risk of being accused of enhancing their electoral fortunes by disproportionately favouring those areas where their support is weak, most frequently by channelling public investment into swing constituencies. The process is known as pork-barrelling and the charge is most frequently levelled at the way Federal spending has been allocated across the USA, though in general it has been shown to be more effective in bolstering the popularity of individual politicians, rather than influencing wider political allegiances.

The electoral map of the UK

At the beginning of the twenty-first century, the electoral map of the UK appears to be experiencing the kind of fundamental political shift described above. The Conservative and Labour Parties have dominated the national political landscape for more than half a century, with long periods of Conservative government being occasionally usurped by radical intrusions of Labour rule, notably between 1945 and 1951, and between 1966 and 1970. After 1979, a string of radical and reforming Conservative governments, led by Margaret Thatcher, seemed be decisively breaking this mould. There was widespread speculation that the Labour Party was finished as a national mass party and that its support base had shrunk to the urban and industrial regions of northern Britain. There was talk of a nation dividing and the prospect of growing social tension between a poorer and politically disaffected north of the country chronically lagging behind a prosperous and modern politically right-wing south. Geographers made some significant contributions to the examination of this growing divide through the analysis and mapping of the changes (Johnston *et al.*, 1988a).

A decade later the electoral map had been transformed. The fortunes of the Conservative and Labour Parties had been almost completely reversed and a third party, the Liberal Democrats, which had been insignificant in electoral terms since the early part of the twentieth century, had made significant inroads into particularly the Conservative vote in Scotland and south-west England. In addition, the nationalist parties in Scotland (Scottish Nationalists) and Wales (Plaid Cymru) had both gained significant numbers of seats in their respective regions and were beginning to resemble more closely the well-established regional parties in Northern Ireland.

In the 1997 General Election the Labour Party made sweeping gains – a net increase of 144 seats (53 per cent) – right across Great Britain, but particularly in the Midlands and south-east England, and in Scotland and Wales (Figures 7.1 and 7.2). The Conservative Party disappeared as a national electoral force in both Scotland and Wales to be replaced by a combination of Liberal Democrats and Scottish and Welsh nationalists. Even in Northern Ireland the traditionally strong links between the Conservative Party and majority Ulster Unionist Party were severely weakened. Aside from the change of government, therefore, the 1997 election seemed to indicate a real resurgence of the Liberal Democrats as a credible national political force with the prospect of it being able to strengthen its position further in the future from a broad base across the UK. The party gained seats in Scotland and Wales, extended its traditional base in south-west and north-west England, and significantly increased its number of seats in the south and east of England. In Scotland and Wales, the nationalist parties made less spectacular gains, but showed their potential for political influence; while in Northern Ireland multiparty politics became a reality in electoral terms with the fragmentation of the once omnipotent Ulster Unionist Party and the emergence of the nationalist Sinn Féin on the national electoral stage in the UK with a second seat in the Westminster Parliament.

The General Election in 2001 confirmed the apparently fundamental shift that had occurred in 1997 (Figure 7.3). The Labour and Conservative Parties altered their relative positions hardly at all. The Conservative Party did make some gains, but these were almost completely offset by losses elsewhere, mostly to the Liberal Democrats. The party did make a solitary gain in Scotland, winning back the geographically large but sparsely populated seat of Dumfries and Galloway from the Scottish Nationalists, but effectively it remained what it had become in 1997, an English right-of-centre party. The Liberal Democrats, on the other hand, further consolidated their spectacular gains in 1997. They added 6 seats to the 28 they had gained then, giving the party 52 seats overall, still only a third of the number held by the Conservative Party, but sufficient for the party to be able to play a significant role in the political opposition. The results for the nationalists were much less encouraging. The Scottish Nationalists lost one seat, while Plaid Cymru lost one and gained one.

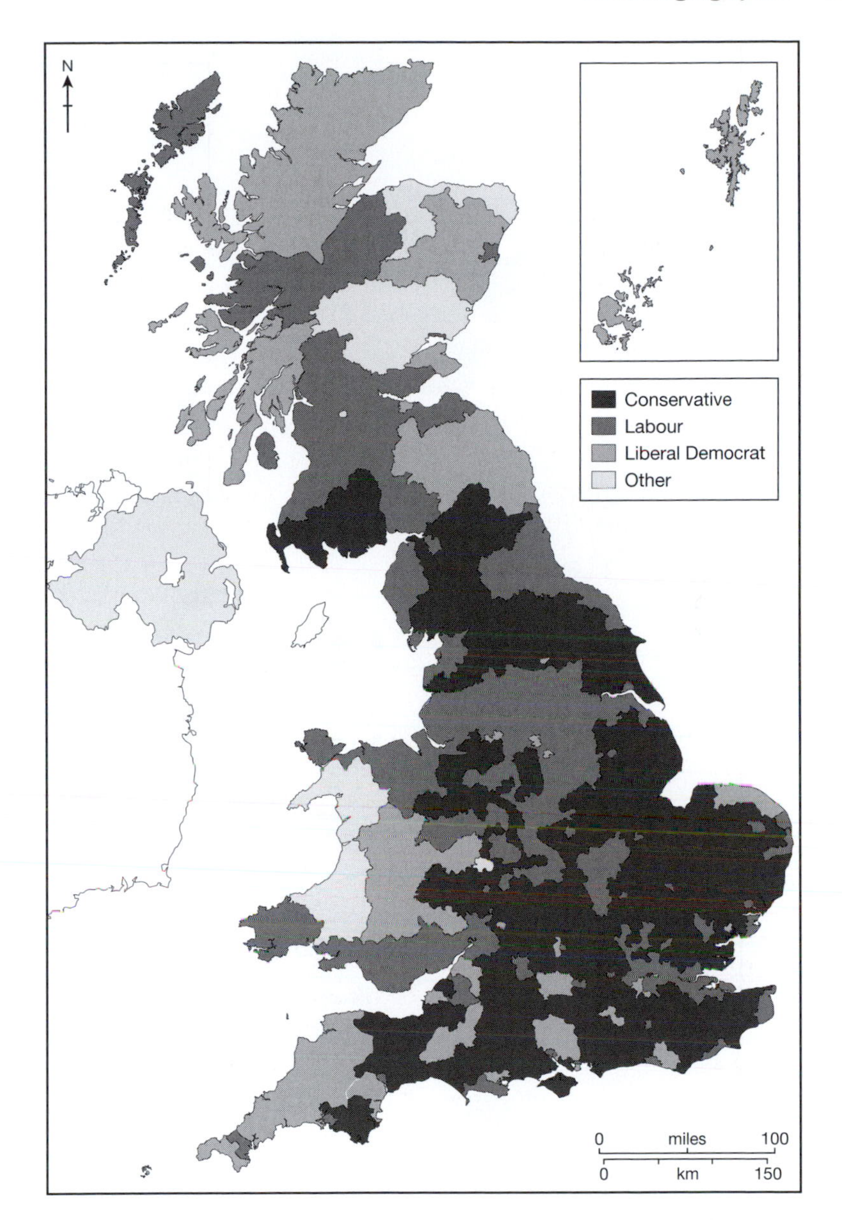

Figure 7.1 ***Electoral map of the UK after the 1997 General Election***

Meanwhile, in Northern Ireland the Ulster Unionist Party continued to cede ground, losing two seats to Sinn Féin and three to the Democratic Unionist Party, making the political situation in the province even more multiparty, but also isolating it to an even greater extent from the rest of the UK.

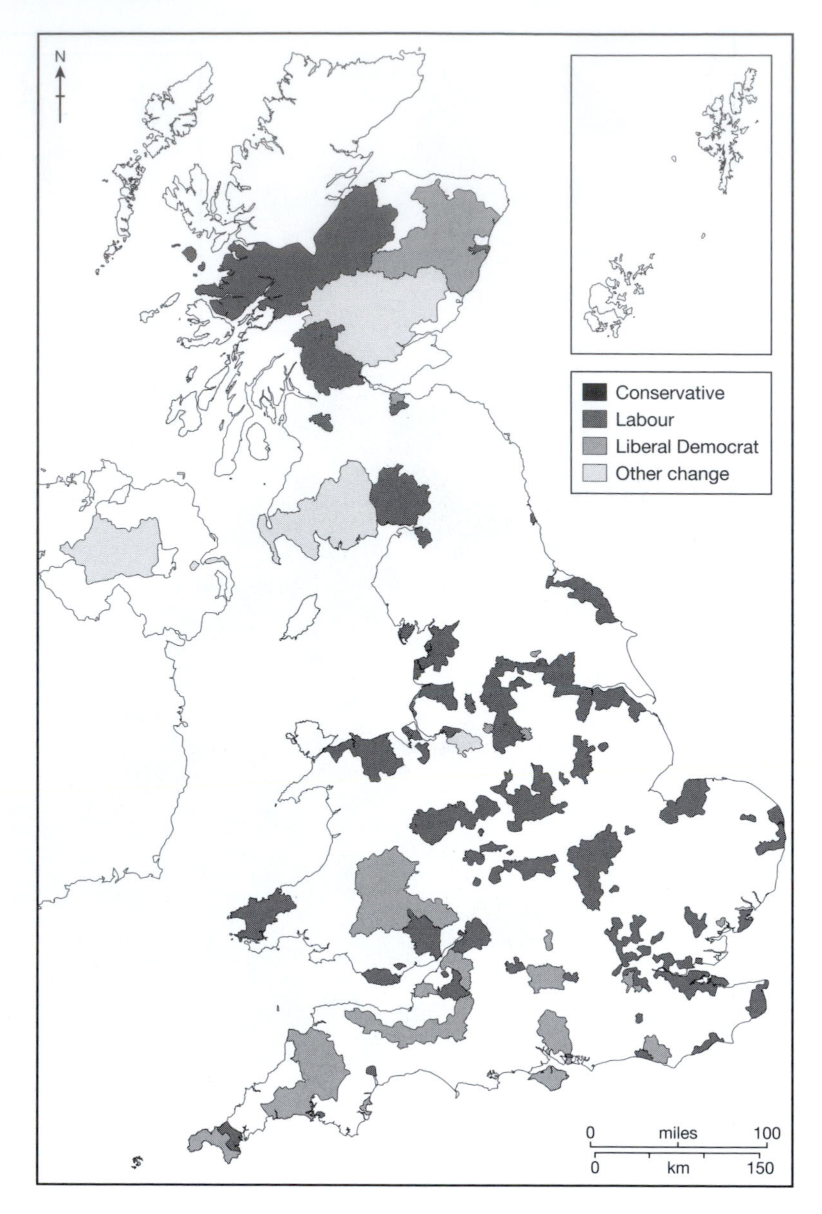

Figure 7.2 ***Changes to the electoral map of the UK after the 1997 General Election***

Predicting trends in political allegiance is naturally an exercise full of danger, but the changes in the electoral geography of the UK at the turn of the twenty-first century seem to show some fundamental changes in the national electoral mood and these changes are clearly illustrated by

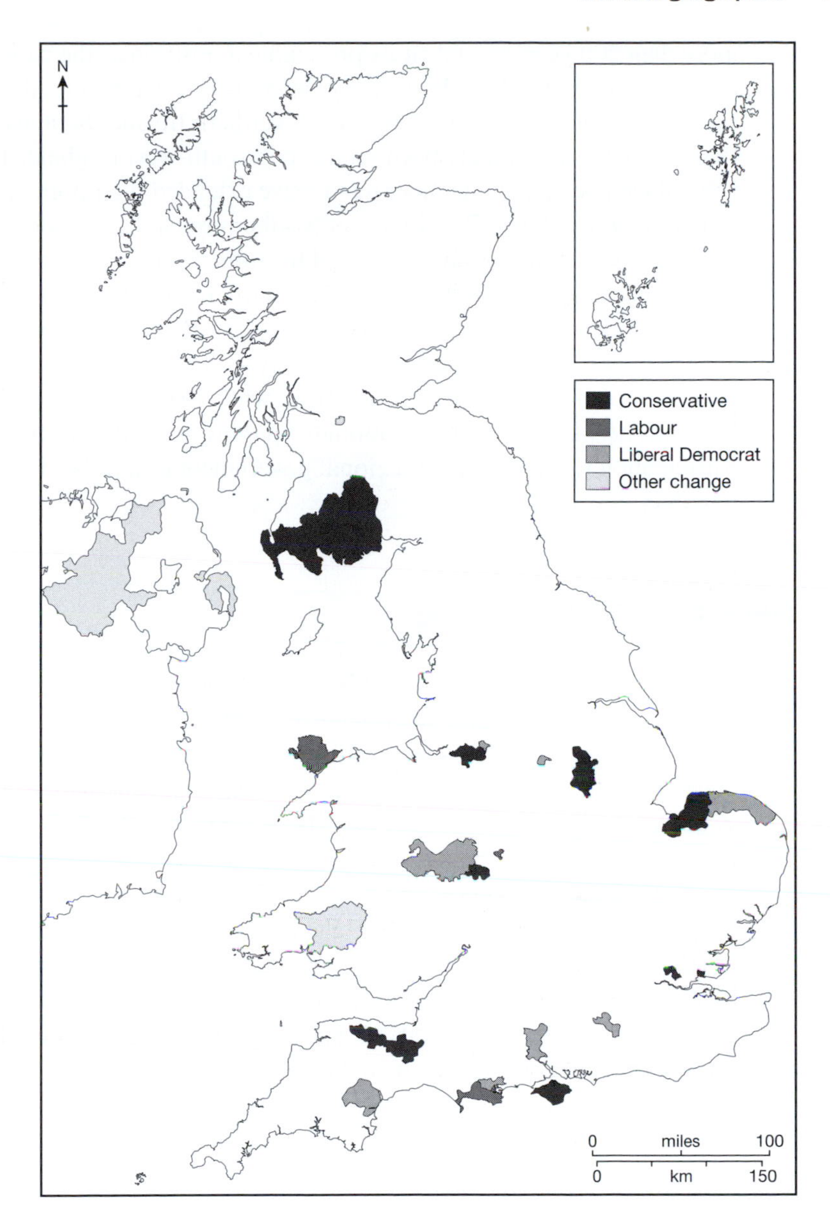

Figure 7.3 ***Changes to the electoral map of the UK after the 2001 General Election***

the cartographic evidence. The Labour Party is now the only political party that can claim to be truly national on the basis of seats won, notwithstanding its complete absence from Northern Ireland, where it has so far chosen not to field candidates. The Conservative Party, which for more than 200 years has taken its pre-eminent position as the national political party in the UK almost for granted, is now barely credible as such. It has shrunk away, not only from Northern Ireland, Scotland, and Wales, but has also largely been replaced in south-west England. The Liberal Democrats, on the other hand, have extended their political foothold all over Great Britain, if not Northern Ireland, and would seem to have a very broad-based launch pad for further growth in their number of parliamentary seats in the future. The outlook for nationalist parties is much less clear. Neither the Scottish Nationalist Party nor Plaid Cymru have really established themselves securely, either on the national stage or in their respective regional assemblies, while in Northern Ireland the divisions between the various nationalist parties remain so deep that participation in any devolved regional government is at present not possible.

Key themes and further reading

At the heart of this chapter is the history of democratic elections and the progress made in widening the franchise in different parts of the world since the beginning of the nineteenth century. It is important to understand the strengths and weaknesses of the main alternative voting systems that are used: a simple majority of votes cast, the single transferable vote, and the list system. The aim of fully democratic elections should be to ensure that each vote cast counts equally towards the election of representatives, but bias is difficult to eliminate. Malapportionment, when there is a variation in the number of electors between voting constituencies, is hard to eradicate and the reasons for this need to be clearly appreciated. Similarly, the processes behind the manipulation of constituency boundaries for political advantage, known as gerrymandering, should be understood. Finally, the reader should have some knowledge of the consequences of long-term shifts in voting patterns at the national level.

Two books were instrumental in propelling electoral geography to the centre of the political geography agenda: *Political, Electoral and Spatial Systems* by R. J. Johnston (1979) and *Seats, Votes and the Spatial Organisation of Elections* by G. Gudgin and P. J. Taylor (1979). They are still the best introductions to the subject written by geographers, although there is a huge amount of material written by political scientists. A comprehensive introduction to the facts and figures of British elections is the book edited by Mike Thrasher and Colin Rallings, *British Electoral Facts 1832–1999*, the sixth edition of which was published in 2000. A fascinating study of the changing electoral geography of the UK is to be found in *A Nation Dividing? The Electoral Map of Great Britain 1979–1987* by R. J. Johnston *et al.* (1988b).

SECTION B

Ideology and geopolitical visions

8 Imagining natural divisions of global power

Map me no maps, sir, my head is a map, a map of the whole world.
(Henry Fielding, *Rape upon Rape*, 1730, Act II, Scene v)

The vision and logic of empire

Geopolitics is a branch of political geography which argues that understanding the dynamics of space is essential for a proper understanding of international relations (Heffernan, 1998). Certainly, manoeuvrings over access to territory have always been an essential backdrop to political strategy and, as Henry Fielding commented, admittedly in a rather different context, grand designs have an enduring fascination. The theme is a universal one and has been a driving force for all the great empire-builders over the millennia. The list of ambitious men – and women are notable by their absence – who have sought to forge empires across what they saw as the known world is almost endless, but a few examples will serve to illustrate the scale and scope of their ambitions.

Alexander the Great, in the third century BC, extended the Greek Hellenic Empire across the whole of the eastern Mediterranean and into Asia Minor and north Africa, thus providing a rich trading hinterland for the Greek city states, which were relatively poor in terms of natural resources. In the third century AD, the Roman Empire, again supporting a massive trading network, stretched from its north Italian core to Britain in the north, across the bulk of what is now known as Western Europe and north Africa, and eastwards into Asia Minor, including the city of Constantinople on the Bosphorus, the modern-day Istanbul. Some four centuries later, in the seventh century AD, the Emperor Charlemagne united the fractious, warring parties in Europe to create an empire that brought together a vast territory, extending west to east from the Atlantic to the Danube, and north to south from the Netherlands to Provence in what is now Mediterranean France. After Charlemagne's death, the eastern part of the empire became the core of the Holy Roman Empire, a loose European confederation with Christianity as its common denominator, that lasted, at least in name, until the beginning of the nineteenth century.

In more modern times, the Emperor Napoleon Bonaparte forged an empire embracing over 40 million people, based on his republican administrative system and centred on France. By 1812, if his allies and members of his family acting as puppet rulers are included, it reached from Spain, across France, to Denmark in the north, to present-day Italy in the south, and eastwards to what is now Germany, western Poland, and Austria. Famously, of course, Napoleon wanted more and he launched an invasion into neighbouring Russia, which ended in a humiliating defeat and ignominious retreat from a point 150 km west of Moscow. The Empire then went into a terminal decline, though its administrative and legal systems proved more durable, surviving essentially little changed to the present day in the independent states that subsequently emerged. The political vision, however, did not entirely die with Napoleon. It was revived in the middle of the twentieth century by the then French President, Charles de Gaulle, who mused about the aim of the European Union being ultimately to create an integrated Europe stretching from the Atlantic to the Urals. Ironically, the collapse of the Soviet Union in 1989 and the extension of the European Union in 2004 to include most of the countries of Central and Eastern Europe has gone some considerable way towards seeing the quasi-Napoleonic vision of de Gaulle become a reality.

By modern-day standards, the extent and ambition of the empires described above are relatively modest, regional rather than global, but the past two hundred years have seen a marked change of scale with empires conceived as part of truly global strategies (Godlewska and Smith, 1994). The legacy of European economic, technological, and military domination was huge empires, some stretching over several continents, dominating great swathes of the world map as described in Chapter 3.

The strategies underpinning the vast majority of these empires were crude and opportunistic, consolidating national control over economic interests worldwide, wherever these happened to lie. They were not thought of in terms of the wider capacity of the imperial home countries to manage and defend the territory claimed. Rather the aim was to achieve as much territorial and economic hegemony as possible and to worry about the detail of what was being claimed once the interior lands had been explored and mapped. The high seas gave free and almost unconstrained access, so that as long as any competing claimants could be kept at bay, there was ample time for leisurely, and profitable, economic exploitation and development.

Geography as an academic discipline, of course, played little part at the time in developing an understanding of the processes at work, other than to spearhead the nineteenth-century exploration of lands previously unknown in Europe (Bell *et al.*, 1994). Nevertheless, the end of the nineteenth century saw growing critical attention being paid to the whole question of imperialism, not least from Marxist theorists such as Lenin, who argued that it was just another phase in the march of capitalism and that either it would bring capitalism to its knees through popular revolution or states would realise just in time that wealth must be redistributed more equitably and take the steps necessary to ensure that this happened. At the time, Geography was just beginning to emerge as a serious and separate academic subject and its distinctive contribution to understanding imperialism was to come on two fronts: first, by seeking to understand better the strategic limitations that would ultimately restrict the scope of global power, and second, by linking competition for power between states to natural laws, in particular those associated with environmental determinism and social Darwinism that were so much in vogue at the time (Peet, 1985) (Box 8.1).

Sir Halford Mackinder and the geographical pivot of history

At the beginning of the twentieth century, geographers across Europe became embroiled in the ferment of ideas about the relative importance of the various factors influencing the distribution of global political power. As mentioned briefly in Chapter 1, the foremost amongst them was the British geographer, Sir Halford Mackinder, who suggested that the pre-eminence of sea power, which had been almost taken for granted for nearly four centuries, since the beginning of the Columbian era, was coming to an end (Mackinder, 1904; Blouet, 1987 and 2004) and being replaced by land power. Central to understanding the new order was the geographical pivot of history, a concept that Mackinder subsequently

Box 8.1

Social Darwinism

Social Darwinism is a loosely defined term, used to describe the application of the evolutionary theory and principles developed by Charles Darwin in the late nineteenth century to socio-economic and political affairs. It has been widely adopted by geographers in their writings, but has come to have largely negative connotations, reflecting the crude competitive values of Victorian capitalism. Descriptive phrases, such as 'the survival of the fittest', have been taken to justify a world where success in the struggle to reach the top of life's greasy pole is the ultimate achievement. At the turn of the twentieth century, however, social Darwinism was for many a liberating philosophy. Much of European society was dominated by class-based hierarchies, that denied many people access to key areas of power and influence. Social Darwinism offered the prospect of access based on achievement and, thus, could be seen as a liberating influence. Nevertheless, it remains a problematic concept, encompassing a wide range of rather simplistic and poorly defined notions about the links between physical evolution and social change.

fleshed out and renamed the Heartland of Eurasia (Mackinder, 1919; Parker, 1982). The Heartland, the core of which was Eastern Europe, roughly coincided with the boundaries of Tsarist Russia and Mackinder confidently claimed that military and political control of this region would assure control of the whole of Eurasia, what he termed the World Island, and that this would make possible global domination (Figure 8.1). He summed up his ideas in what has become a very hackneyed jingle, though none the less powerful for that:

> Who rules East Europe commands the Heartland;
> Who rules the Heartland commands the World Island;
> Who rules the World Island commands the World.
>
> (Mackinder, 1919)

It was a very deterministic view of the world and Mackinder was no doubt influenced by the vogue for environmental determinism in the late nineteenth and early twentieth centuries, as well as by the fascination amongst geographers at the time with the identification of so-called natural regions (Box 8.2). This enthusiasm was particularly marked in the School of Geography at Oxford University under the influence of A. J. Herbertson, where Mackinder also taught at the time (Herbertson

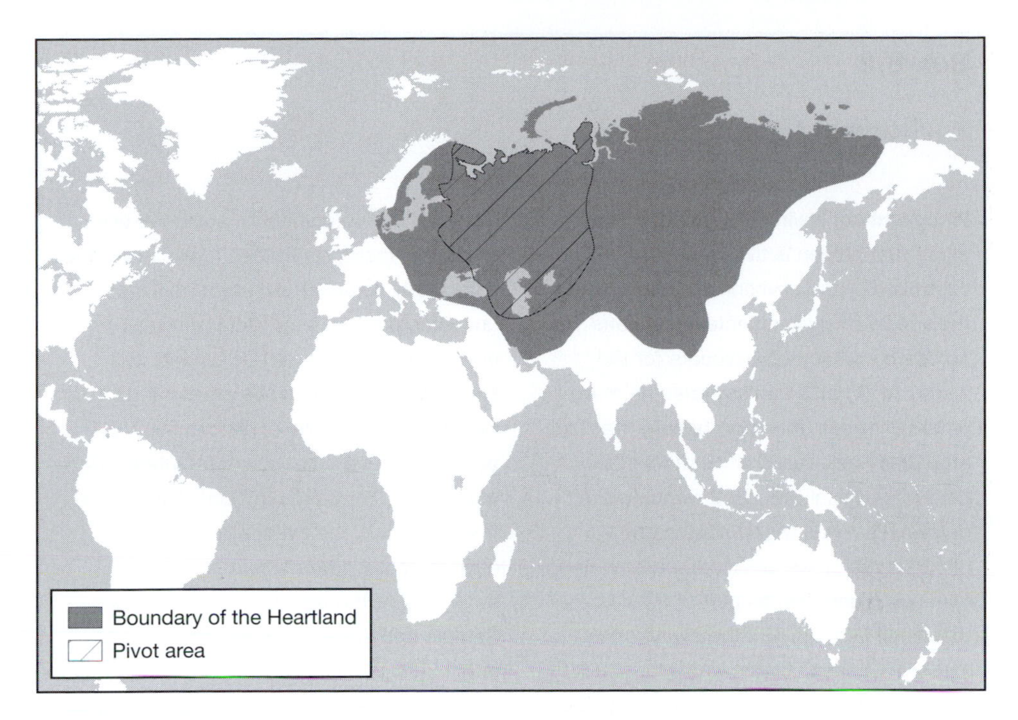

Figure 8.1 ***Heartland/World Island***

and Herbertson, 1899). With the gift of hindsight, it now seems a somewhat absurdly simplistic model for explaining something as complex as the emerging world order at the height of the industrial revolution, but it was a time when the major industrial nations were still engaged in a period of rampant imperial expansion and it very much caught the mood of the times. The concept was also actually extremely vague, as is only too evident from the crude published sketch maps, but it was to prove extremely politically influential in the period between the two world wars, something that will be analysed in more detail in Chapter 9.

By no means everyone was convinced by Mackinder's thesis, especially in the USA, the emerging power of which, it can now be seen from the pre-eminent position of the USA in world affairs, was chronically underestimated in the analysis. The American naval historian, Admiral Alfred T. Mahan, who had already written extensively on the role of sea power, was immediately very sceptical of the way Mackinder sought to downgrade its importance, a view subsequently developed in more detail by the American Nicholas Spykman (1938 and 1942).

Box 8.2

Environmental determinism

Proponents of environmental determinism argue that human activities are ultimately controlled by the environment and that it is the variety of environmental conditions across the world that accounts for the variety of peoples and societies. In many different guises, it is a philosophy that has repeatedly resurfaced since at least the time of classical Greece. Environmental determinism was widely adopted by geographers in the late nineteenth and early twentieth centuries, as part of their extensive flirtation with the evolutionary ideas of Charles Darwin, which he set out most notably in the *Origin of Species*. They attempted to use the concept to explain the differential development of societies across the world, though in Europe it was never the completely dominant paradigm, nor was it uncontested. For many, determinism, environmental or any other variant, was simply too dogmatic a stance and it had to vie with the more liberal philosophies of possibilism and probabilism for intellectual dominance. For a time in North America, in the first half of the twentieth century, environmental determinism held greater sway, largely through the writings of Ellen Semple and Ellsworth Huntington, but its intellectual dominance there was short-lived, not least because of the racist conclusions the philosophy frequently spawned.

Spykman proposed an alternative vision, based around what he termed the Rimland, a buffer zone between land and sea, especially in the Indian and western Pacific Oceans (Figure 8.2). The effect of the Rimland buffer zone would be to contain the rampant land power that Mackinder believed would become so dominant and all-pervasive. Although both visions were extremely speculative, Spykman is no less important than Mackinder, not least because US foreign policy strategists enthusiastically adapted his ideas to help realise their goal of containing the global Communist threat in the Cold War era after the Second World War.

By the middle of the twentieth century, global hypothesising about power relations had become largely discredited and fallen out of fashion in geography, but nevertheless refused to die and disappear completely. In the context of the Cold War, Saul B. Cohen proposed a system of geostrategic regions, a two-fold division into what he termed the Trade-Dependent Maritime World and the Eurasian Continental World (Cohen, 1964 and 1982). The former included Western Europe, the

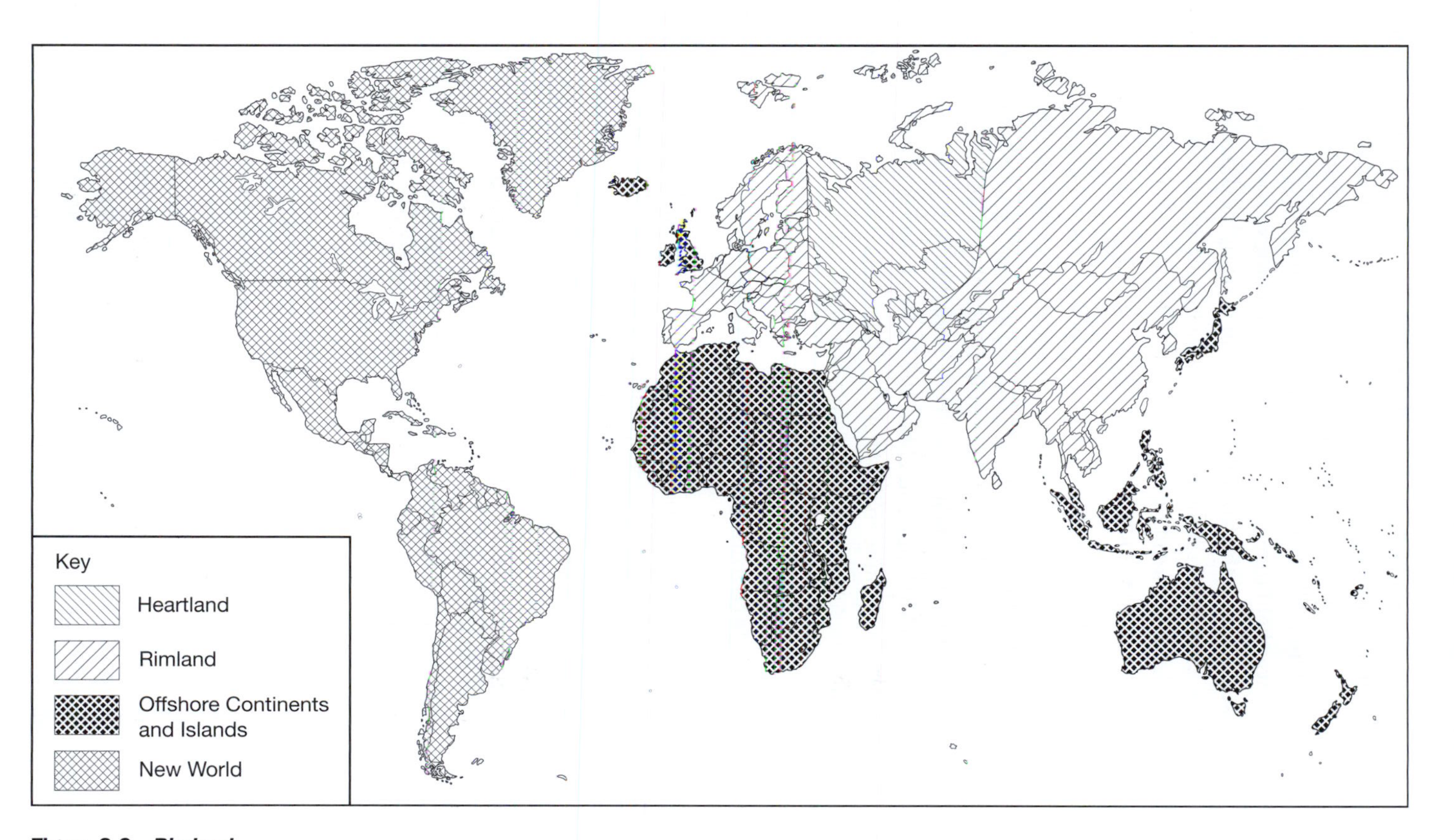

Figure 8.2 ***Rimland***

Americas, most of Africa, and Australasia and was bound together by its commitment to free trade through a complex system of maritime trading links. The latter included the Communist World centred on the huge land-based block of the Soviet Union and the People's Republic of China and was held together by Communist ideology. The model was a restatement of the Cold War divide in geographical parlance and has had to be completely rethought since the collapse of the Soviet Union in 1989 and the subsequent rapid expansion of trade between China and the rest of the world (Cohen, 1992).

The current preference for trying to make sense of the spatial structure of the world order is through the emergent trading blocs. The most successful thus far have been the EU and NAFTA, the North American Free Trade Association, but similar, though less ambitious and successful, experiments have been undertaken in Asia, Africa, and South America (Michalak and Gibb, 1997). It is a decisive departure, being rooted much more in the analysis of actual economic flows, rather than being a means of achieving a particular political view of the world.

Environmental determinism and the state

The influence of environmental determinism and social Darwinism on intellectual life in Europe, North America, and the English-speaking world generally at the end of the nineteenth and the first half of the twentieth centuries was profound and widespread. Many geographers at the time absorbed its messages and the impact on their subsequent writings was clearly evident. One of the most important of these people was Friedrich Ratzel, whose book, *Politische Geographie* (1897), for the first time identified political geography as a distinct subdiscipline and also laid it firmly within the context of the debates about environmental determinism that were current at the time.

The most important of his arguments were those relating to the nature of nations. He compared their growth to that of living organisms, which evolved and grew over time, making the fairly obvious point that as nations grew, they would inevitably outgrow their state boundaries and require more resources, leading to competition and conflict. Amongst these resources, space was identified as the most important, and it was at this point that the concept of *Lebensraum* – living space – was born. Ratzel depicted nations as being in an endless struggle to dominate space and to acquire enough of it to be able to live and breathe, a struggle in which there were inevitably winners and losers. It was certainly far removed from the ordered mosaic of states, fixed in time, that the static picture in atlases portrayed. However, it is important to note that Ratzel himself recognised the limits of the biological analogy he had used, freely admitting that political and cultural constraints would inevitably influence how and when boundaries changed (Bassin, 1987b).

One of the most intriguing aspects of Ratzel's work is the extent to which he was influenced by the American historian Frederic Jackson Turner. Writing just three years before *Politische Geographie* was published, Turner (1894) propounded the frontier thesis, which saw the settlement of the United States as a dynamic process with a series of waves moving inexorably across the North American continent from east to west, but with each temporarily checked by a marked boundary – the Appalachian Mountains, the Mississippi and Missouri rivers, and the Rocky Mountains – and a series of Indian wars, before settlement finally reached the Pacific west coast (Block, 1980).

Turner's thesis was widely debated at the time and its impact was certainly not confined to the United States (Kearns, 1984). Indeed, it spawned a number of disciples, such as W. P. Webb, who used it to explain the course of European settlement across southern Africa and Australia and other parts of the world allegedly being civilised and brought into the purview of the modern world. Whatever the extent of its influence, Turner's thesis has been roundly criticised and dismissed in recent years (Limerick, 1987). Not only did it not fit the facts of the history of settlement across the United States, which was much more fractured and irregular than Turner had described, but the idea that the West consisted entirely of free land, there to be exploited by the European settlers, buried the legacy of thousands of years of a quite different form of settlement by native peoples which was culturally highly sophisticated, though pre-industrial.

Ratzel was undoubtedly aware of the frontier thesis, though he does not make any explicit reference to it. However, whereas Turner clearly misinterpreted the facts of North American European settlement history, Ratzel himself was in all probability misrepresented by many of those who subsequently used his work to try to justify conquest under a cloak of seeking adequate *Lebensraum* for rapacious nations. His work must always be viewed in the context of when it was written, and the late nineteenth and early twentieth centuries was a period when determinism, including environmental determinism, was the dominant paradigm. In the subsequent century there have been several important paradigm shifts and the development of society is now seen in a much more varied and flexible light (Kuhn, 1970; Johnston, 1991).

Frontiers and boundaries and the science of geopolitics

One of the main consequences of so much of European and American intellectual life being dominated by environmental determinism in the early years of the twentieth century was what amounted to an obsession with frontiers and boundaries (Prescott, 1987). Peter Taylor concluded that: 'Frontiers and boundaries have probably been the most popular topic in political geography' (1985, p. 104) and it is important to understand the reasons behind their enduring popularity (Box 8.3). The key lies in the general preoccupation amongst geographers with the search for natural regions in the early part of the twentieth century, especially in the Anglo-American literature (Fleure, 1919; Hartshorne, 1939 and 1958). The idea that there were such spaces as natural regions led logically to a belief in natural boundaries, which once identified could be turned into political realities and defended. From this position, it was but a small step to arguing the case for changes to existing political boundaries that failed to meet the criteria of natural regions. Nor was it

Box 8.3

Frontiers and boundaries

The terms 'frontier' and 'boundary' have been widely used, often interchangeably, by geographers and others to describe political divisions, especially at national level (Prescott, 1987). Given this background, it is difficult to draw a clear distinction between them. At the height of the European imperial expansion in the nineteenth and early twentieth centuries, frontiers were a common part of the political parlance, representing either the political division between two states, or the limit of permanent settlement. Today, frontiers in that latter sense have all but disappeared under the global tide of human settlement and economic development, to be replaced by boundaries, which are more unequivocally lines of demarcation. Although both terms are still used in geography, the word 'boundary' is clearly the accepted generic term for describing political divisions.

illogical to claim that as these were partly human constructs, as demographic and other social constraints changed, so should frontier and boundary lines.

Politically, the timing was opportune. At the beginning of the twentieth century, imperialism was at its height and there was a pressing need amongst the European colonial powers to devise defensible systems for defining politically the huge tracts of territory, particularly in Africa and Asia, to which they had laid claim. The best known and most influential of the exponents of the science of boundaries in the UK was Lord Curzon, a politician and polymath who was appointed Viceroy of India in 1899 and Foreign Secretary in 1919, as well as serving as President of the Royal Geographical Society. In 1907 he delivered the Romanes Lectures in Oxford on the subject of frontiers, in which he made a strong plea for frontiers and boundaries to be scientifically determined, even allowing for the fact that it was unlikely that any single set of principles for doing so would ever be agreed (Curzon, 1907). His enthusiasm for such general principles was highly empirical. There was evidence all around him that existing frontiers and boundaries had failed to bring political stability and he believed that a better organised political map would allow for much greater success. Interestingly, the horrors of the First World War subsequently made him much more cautious and, once he was involved as the British Foreign Secretary in devising a post-war settlement, he clearly understood that political horse-trading would be as important as any set of scientific principles.

Elsewhere in Europe the fascination with boundaries was equally great. In Germany, Albrecht Penk, best known now for his contributions to physical geography, gave his inaugural address as Rektor of the Friedrich-Wilhelms Universität in Berlin on the subject of political boundaries (Penk, 1917). In it he extolled the sacrifices being made by Germany's youth to defend the borders of the German Empire, but stressed that he viewed Germany's colonial boundaries more as areas of contact than friction, facilitating peaceful interchange between neighbouring peoples.

Others in Germany took a more robust approach and, after the First World War, developed a school of political geography, known as *Geopolitik*, which was explicitly influenced by the earlier writing of Friedrich Ratzel on the organic theory of the state. In fact, some of those who initially took up Ratzel's ideas, in particular the Swedish political scientist Rudolf Kjellen, the first person to coin the word *Geopolitik*, probably exerted even greater sway. The leader of the German school was Karl Haushofer. Largely through the academic journal, *Zeitschrift für Geopolitik*, of which he was the editor, he and his collaborators generated a large literature, geared primarily towards rehabilitating Germany as a major European power in the wake of the defeat of the Kaiser's Second Empire in 1919 (Sander and Rossler, 1994). Haushofer's massive three-volume work, *Macht und Erde* (Power and Land), set out a detailed case for a radical redrawing of political boundaries across the globe to bring them more into line with what he saw as geopolitical reality (Haushofer, 1931–4). It evoked an immediate reaction in the English-speaking world, notably from the American geographer Derwent Whittlesey, who argued for a much more subtle interpretation of the relationship between society and the land. Rather than the natural environment being automatically the dominant influence in determining the most appropriate political boundaries, Whittlesey argued that they should emerge as the result of the long-term influence of human activity on the land (Whittlesey, 1939; Cohen, 2002).

Much of what was written in the name of *Geopolitik* was crude and self-serving, such as the tract *Spaniens Tor zum Mittelmeer* (Spain's gateway to the Mediterranean), which supported and justified Spain having exclusive control over the land either side of the Straits of Gibraltar (Pauser, 1938). It and other publications in similar vein were widely vilified at the time outside Germany, but nonetheless had the effect of discrediting worldwide the whole tradition of geopolitical analysis and writing for much of the second half of the twentieth century, including any suggestion that boundaries and frontiers could legitimately be seen as dynamic and changing elements in the landscape.

Geopolitics and the scientific study of boundaries was a tradition by no means confined to Germany and many of its exponents in other countries in Europe argued strongly against using boundary issues in any way as a justification for war, as opposed to the promotion of peace and harmony between nations. Geographers, such as Elisée Reclus in France and the Russian anarchist Peter Kropotkin writing in exile in the USA, provided substantial counter-arguments to those of *Geopolitik*, but their arguments did not attract the same level of public interest at the time (O'Loughlin and Heske, 1991).

The most effective counter to what he himself described as 'the geopolitics of domination' has come only relatively recently in the writings of Geoffrey Parker (1988). He suggests that this is only one of two main traditions in geopolitics, the other being the geopolitics of cooperation. Although the latter has attracted much less serious academic study by political geographers, its achievements in practice throughout the twentieth and into the twenty-first centuries have been considerable. The peace initiatives throughout the period between the two world wars, as well as the doomed attempts to create an international political forum in the League of Nations, were not the end of the process of creating a world order based on peaceful coexistence. Quite the reverse, they marked the beginning of a major shift in geopolitical practice that, though interrupted between 1939 and 1945 by the Second World War, has continued to gather momentum through bodies such as the United Nations and the European Union.

Despite the now virtually universal rejection of the rather crude environmental determinism associated with the search for natural boundaries, boundaries and their delimitation continue to exert a powerful fascination for geographers, not to mention politicians. There are several major academic research institutes in the world devoted entirely to the study of frontiers and boundaries, including two in the UK (International Boundaries Research Unit at the University of Durham and the Centre for International Boundaries Research at the Queen's University Belfast), an indication of the continuing level of activity and its importance to political geography.

Critical geopolitics

The dogmatism and prescription associated with the study of geopolitics, and the way in which it had been used to promote particular national territorial agendas, particularly in Germany during the Third Reich (1933–45), not only brought this aspect of political geography into disrepute, it also meant that it was ignored by many political geographers. Geopolitics desperately needed to be reassessed and reinvigorated, but it was not until the concepts of postmodernism and post-colonialism began to interest geographers that there was sufficient impetus to move the subject forward (Dalby, 1990; Driver, 1992).

Many of the ideas for this reappraisal originated from outside geography, a notable source being the writings of the Palestinian cultural historian Edward Said, who argued forcefully in his book *Orientalism* that moral right in the world was rarely if ever vested in one side only. The views of what he termed 'the Other' were always relevant and that by looking at political issues from more than one side, and especially from a multicultural perspective, was likely in the long run to lead to more enduring solutions (Said, 1995).

The lessons were embraced enthusiastically by political geographers, eager to break out of the philosophical straitjacket they had been struggling with for so long (Ó'Tuathail, 1996). The main contribution of critical geopolitics has been to broaden greatly the focus of debate within geography about relationships between states and other political entities. A seminal study examines the changing nature of relationships along the border between Finland and Russia, which have been subjected to a fundamental process of 'deterritorialisation' and 'reterritorialisation' (Paasi, 1996). What is implied by this rather abstruse formulation is that political relationships rarely remain stable for long; there is always an ongoing process of questioning and reassessment that results in the old order changing and new ones replacing it. After the fall of the Soviet Union in 1990, this process was particularly active along the Finnish–Russian border, leading politicians and peoples on each side to reassess each other's motives and aspirations. In practical terms, what had been viewed as a virtually impermeable front line, became almost overnight a zone of fruitful and mutually beneficial contact. The intellectual energy generated by critical geopolitics is undeniable, but is more difficult to categorise and summarise within a neat framework. Very broadly, however, there are three basic organising concepts. First, there is politics associated with all types of geographical knowledge; second, there is a geography to all political practice; and third, the first two ideas can only be uncovered by challenging the taken-for-granted (Taylor, 2000, p. 126). In other words, there is no such thing as a value-free political decision and all political decisions have spatial consequences. One of the most important roles of political geography is to challenge the authority of decisions about territory and boundaries, so as to uncover their total impact on all those affected and, thus, help counter a one-sided and partial interpretation of events. It is a very different role from that envisaged by Mackinder and Ratzel.

Key themes and further reading

This chapter is about geopolitics and its ramifications for geography. An appreciation of the nature, historical extent, and human ambition behind some major past empires provides an essential background to the concept. The enduring importance of the vision and world view of Sir Halford Mackinder must be understood, in particular the thinking underpinning the geographical pivot of history and the idea of the Heartland. Other, alternative world views, notably those of Mahan, Spykman, and Cohen, should also be included. The influence of environmental determinism and social Darwinism on the way in which many of these ideas evolved, as well as their impact on the dynamics of states, should be clear. The contribution of Friedrich Ratzel and the concepts of *Lebensraum* and the organic theory of the state are important in this context, as is the frontier thesis of Frederick Jackson Turner. The consequences of the heavy criticism heaped upon the whole study of geopolitics, as a result of its self-serving application in Germany through the *Geopolitik* movement, must be appreciated. At this stage the reader should be able to explain why frontiers and boundaries have been so important in the development of political geography, and why critical geopolitics has had such a transformative impact.

There is no better starting point for finding out more about the mindset behind early geostrategic thinking than Brian Blouet's (1987) biography of Sir Halford Mackinder. The tumultuous political upheavals of the twentieth century have meant that geostrategy and geopolitics have exerted a continuing fascination for geographers and Peter Taylor (1993) has edited an excellent collection of essays on the changes over the past century in *Political Geography of the Twentieth Century: a global analysis*. The recent emergence of a more critical approach to geopolitics, which has exposed the limitations of many of the earlier, more dogmatic, approaches, is well outlined by Gerard Ó Tuathail (1996) in *Critical Geopolitics*.

9 Dreams into action

The making of national foreign policy

How is the Empire?

(Last words of King George V, *The Times*, 21 January 1936)

The lure of empire

Although the word geopolitics was only coined in the early years of the twentieth century, applied geopolitics has been practised from time immemorial by those seeking to extend their political power and influence across the globe. The academic interest by geographers in the whole concept of geopolitics in the early nineteenth century led quickly to their being co-opted by policy-makers in government to help them develop and justify new strategies for reforming and stabilising the world order. It was a heady period of public recognition for the discipline, which was just beginning to establish itself as a serious area of academic study in its own right. Unfortunately, as was explained in Chapter 8, the love affair was short-lived. Geography and geographers were accused of providing specious scientific and academic justification for political strategies that were no different from those of a succession of rapacious political and military leaders through history, who had sought to extend their territorial power and influence at the expense of their weaker neighbours.

Ever since being used in this way, geopolitics as a legitimate area of academic study has been treated with understandable caution. At the same time, however, the word itself has become lodged in the wider consciousness and political strategists have incorporated it into their general vocabulary, to a point where geopolitics has come to be seen as a legitimate, and integral, part of their trade. The word is now used as a catch-all term to describe the whole process of trying to manage global political events for narrow, and usually nationalist, strategic ends (Dodds, 2000; Taylor, 1993).

There is no single, simple, explanation of the urge to create global empires, but their importance for the wielding of political power is beyond doubt, as the last words of the dying King George V amply demonstrate. From the earliest times, the scale and ambition of the enterprises is quite breathtaking. Trade is undoubtedly one very important factor and some of the most extensive imperial projects have been built on the back of successful trading networks. The British Empire in the seventeenth and eighteenth centuries owed much to the initiative of trading companies, like the British East India Company (founded by Royal Charter in 1600), the London Virginia Company (1606), and the Hudson's Bay Company (1670). There were also similar companies founded in other European imperial countries, notably the Dutch East India Company and the Dutch West India Company in the Netherlands (founded in 1602 and 1621 respectively), the French East India Company (founded in 1664), and the Swedish East India Company (founded in 1731). All played vital roles in securing the political power bases overseas of their sponsoring governments.

The British East India Company

The most successful and influential of these companies was undoubtedly the British East India Company and the history of its development illustrates graphically how trade and politics were intimately combined in a geopolitical strategy, stretching over more than two and a half centuries (Lawson, 1993). The company was founded by a group of merchants in London in 1600 and was initially granted a Crown charter by Elizabeth I to trade in the East Indies for a period of fifteen years. The East Indies was an indeterminate area, at the limit of the known world as far as Europeans were concerned, but the company succeeded in establishing a number of highly profitable trading posts on the eastern seaboard of the Indian subcontinent. Its success persuaded James I to renew the charter indefinitely in 1609, though with commendable caution he included a clause to the effect that it would be automatically rescinded if the trade turned out to be unprofitable for three consecutive years.

The company expanded quickly and by 1647 had twenty-three factories scattered along both the east and west coasts of India, including Chennai (Madras), Mumbai (Bombay), and Kolkata (Calcutta), all of which were destined to become major cities in modern-day India. The most important of these factories in time became walled forts, such as Fort William in Bengal, Fort St George in Madras, and Bombay Castle, and acted as major focuses for the subsequent urban expansion. The military defences were necessary to ward off other mercantile adventurers, in particular those from the Netherlands, Portugal, and France, all of whom were equally keen to obtain a slice of the lucrative trade in raw materials, such as cotton, silk, indigo, saltpetre, and tea.

In India itself, the Mogul Emperor in Bengal, and other regional rulers, were delighted with the burgeoning trade with England and the prosperity the British East India Company brought with it. They did everything they could to encourage its further expansion, including almost completely waiving all customs duties. This boundless trading success also delighted the government in Britain and in a series of acts King Charles II gave the company the right to annex territory in his name, to mint money, to command fortresses and troops, to forge alliances and make war and peace, and to exercise both civil and criminal jurisdiction over its territories. In short, the company had assumed virtually the full panoply of state powers (Wild, 2000).

Following its overwhelming success in India, the British East India Company was still eager to open up new markets elsewhere in the world and cast envious eyes at the Dutch spice trade in the Far East. By the early eighteenth century, it had secured a strong foothold in the Malacca Strait in what is now Singapore, and also established a trading post at Guangzhou (Canton) and founded what was to become Hong Kong. The company became by far the largest element in the emergent British global market, giving it unparalleled influence on, not only overseas trade, but all areas of government policy.

It was in India, however, that the company really evolved into a state within a state. Between 1754 and 1763, the company and its military forces conducted an extended campaign against the French in India, aimed at removing them as meaningful trading competitors. This culminated in a famous victory by Robert Clive, a company employee and Governor of Bengal, which saw the Fort of St George in Chennai (Madras) recaptured and the French presence in India reduced to a few small coastal enclaves with no military support.

Despite the grip that the British East India Company had over the Indian subcontinent, by the end of the eighteenth century it was clear that it was becoming hugely overstretched trying to cope with all its administrative, judicial, and military responsibilities. The British government was forced to intervene and, in 1773, passed the Regulating Act for India, which introduced sweeping reforms, including specifying the respective roles and responsibilities of the government and the company. For the first time, a Governor General was appointed, independent of the company and directly responsible to the government in Britain. Essentially the company was allowed to retain its virtual monopoly on trade in return for paying all the administrative costs of the British presence in India, but the arrangement proved disastrous. There was still a lack of clarity about the division of responsibilities; the company continued to lose money; and the first Governor General, Warren Hastings, was indicted for corruption, recalled to Britain, and impeached by Parliament.

Subsequently, the arrangement struggled on, but the role of the company and its relationship with the British government became increasingly anomalous, even though its rule extended over most of India and Burma and also included Singapore and Hong Kong, incorporating more than 20 per cent of the world's people. The beginning of the end was when it was deprived of its trade monopoly in 1813 and, after the Sepoy Mutiny, a popular uprising by the Indian peasantry in 1857, the company's official activities in India were finally wound up in 1858.

In terms of geopolitics, the story of the British East India Company is very instructive. Its two hundred and fifty-year history in India spanned virtually the whole of what became known as the age of mercantilism, a time when governments in Europe used overseas trade as a means of extending their global political power and influence (Lipson, 1956). By the end of the eighteenth century, however, it was becoming clear that governments with territorial ambitions could no longer pursue them under the cloak of trading companies. Rather, they had to be more directly involved themselves and create the conditions under which trade could then flourish. It was a major shift of emphasis and marked the beginning of an era when governments began to work much more directly with each other in their far-flung colonial land dealings.

The birth of modern geopolitics

As outlined in Chapter 3, the modern political map, in the form that we know it today, and the idea that it could be manipulated and managed through diplomacy, first began to emerge, very tentatively, in the wake of the final collapse of the Napoleonic Empire in Europe in 1815. Two developments on either side of the Atlantic Ocean were the touchstones for the putative new order: in Europe, the Congress of Vienna, first convened in 1814, began the long process of trying to rebuild the political infrastructure of the continent after a generation of war; while in America the proclamation of the Monroe Doctrine in 1823 by the USA served notice to the European colonial powers that their days of hegemony on the other side of the Atlantic were coming to an end.

The Congress of Vienna

In 1814, Napoleon was apparently finally defeated in France, the monarchy in the person of Louis XVIII was restored, and the Treaty of Paris ending hostilities was signed. Nevertheless, Europe was still in political chaos and, in an ambitious and flamboyant attempt to try to bring back a semblance of order, the Austrian Emperor, Francis I, convened the Congress of Vienna. The other major European economic and military powers, Russia, Prussia, and Britain were all represented, as well as all the multitude of states, large and small, that had existed prior to Napoleon's military and political intervention.

The deliberations at the congress were self-indulgent, slow, and cumbersome, with most of the real negotiations taking place behind the scenes between the representatives of the four major powers. The air of smugness was rudely shattered, however, by Napoleon's escape from Elba in 1815 and his remarkable reassertion of his political and military authority. For a hundred days he marched across Europe gathering support as he went, only to be defeated at Waterloo, after a very close-fought battle.

The effect on the Congress of Vienna was electrifying. It now had a real sense of urgency and a whole host of decisions were quickly taken, which completely redrew large parts of the political map of Europe and its overseas territories (Figure 9.1). Louis XVIII in *France* and Ferdinand VII in *Spain* were confirmed respectively as the monarchs in these two countries. The *German Confederation* of thirty-nine states replaced the several hundred mini-states, dukedoms, princedoms, and the like, that were the last vestiges of the Holy Roman Empire in mainland Europe. This left *Prussia*, which was part of the Confederation, without the prizes it had hoped for: Alsace and Lorraine in the west and Warsaw in the east. By way of appeasement it was given half of Saxony instead, in addition to important parts of Western Pomerania, Westphalia, and Rhine

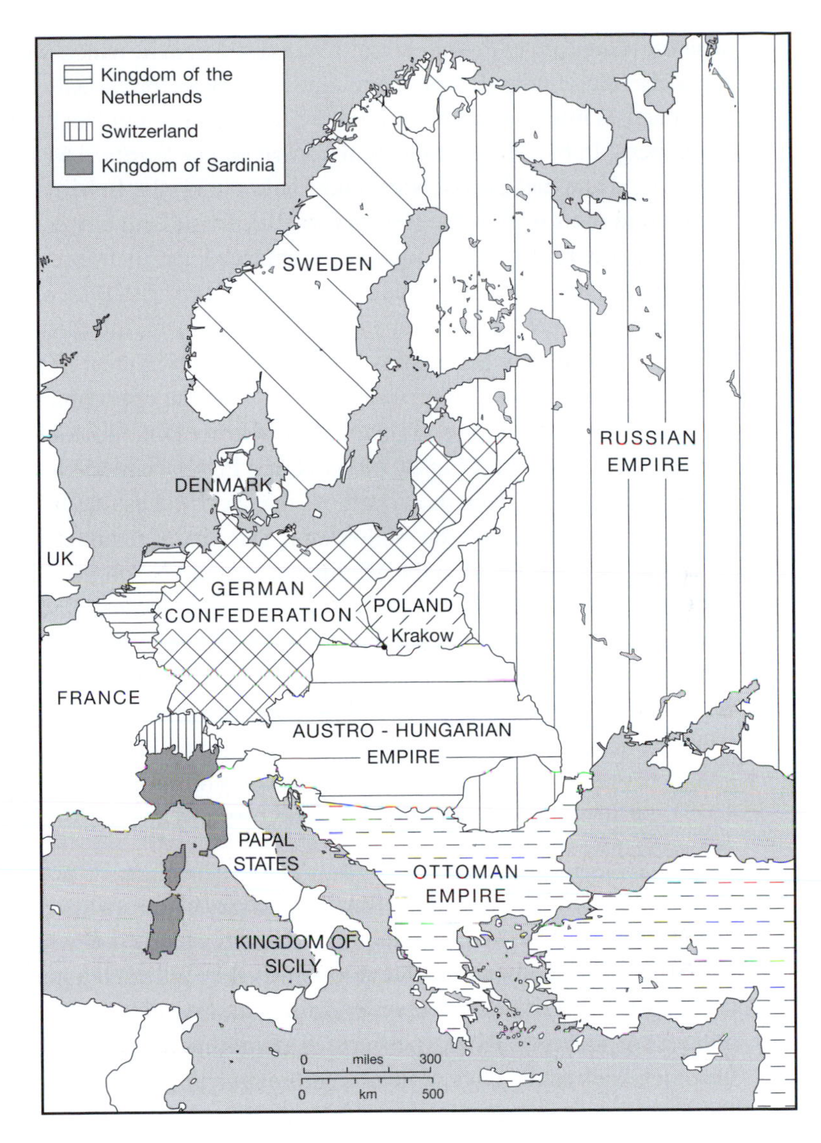

Figure 9.1 ***Territorial changes as a result of the Congress of Vienna, 1814–15***

Province. In similar vein, *Austria* lost its stake in the Netherlands and was compensated with much of northern Italy, as well as the Venetian part of Dalmatia on the coast of the Adriatic. Combining the former Austrian territory in the Netherlands with the United Provinces in the north created the *Kingdom of the Netherlands* under the House of Orange. Russia had its occupation of Finland, Lithuania, and eastern Poland confirmed, and the Tsar was also crowned king of a nominally separate *Kingdom of Poland* around Warsaw. *Sweden*, which had formerly ruled Finland, was given instead Norway, previously ruled by Denmark. While *Denmark* received the former Prussian Duchy of Lauenberg. In the Mediterranean, the *Kingdom of Sardinia* was resurrected and had Savoy, Nice, and Piedmont on the Italian mainland returned to it, as well as the territory of Liguria and the city state of Genoa. *Switzerland* had its boundaries enlarged and its future neutrality confirmed. The southern German provinces of Bavaria, Baden, and Württemberg were all awarded additional territory giving them more or less their present-day boundaries. *Britain* had little interest in acquiring territory on mainland Europe, preferring to take the opportunity to expand its colonial empire. It retained the former Dutch colonies of Ceylon (Sri Lanka) and Cape Colony (Cape Province in the Republic of South Africa), was given those parts of the West Indies that had previously been ruled by the Netherlands and Spain, retained the islands of Malta and Heligoland in the Mediterranean and North Seas respectively, and was given a protectorate over the Ionian islands in the Ionian Sea. All the regimes supported by the congress were monarchies of some sort, with one small exception. The city of *Krakow* in what is now south-western Poland was claimed by Prussia, Russia, and Austria without success and it was constituted as a republic. The final act of the congress was to conclude a new peace treaty with France (Treaty of Paris 1815), confirming its new boundaries and formally readmitting it as an independent state in its own right.

By any standards the wholesale territorial adjustments achieved by the Congress of Vienna were astounding and unprecedented, but opinion is sharply divided about the extent to which they reflected a genuinely thought-through new political order, or just an extended exercise in political expediency. Its sharpest critics claim that it did nothing but turn the clock back to the days of the *ancien régime* in pre-revolutionary, and pre-Napoleonic, Europe (Davies, 1996, p. 762). Others are more generous, arguing that it heralded a period of rapid political, social, and economic change in Europe, without the accompanying curse of war (Nicholson, 1946). Whatever the disagreements, however, the congress did inaugurate a system of regular consultation through international congresses for trying to find diplomatic resolutions for disputes between sovereign states. It is a process that was continued intermittently throughout the nineteenth century with limited success, gathered momentum and general acceptance in the twentieth century, and is now generally the norm for international negotiation.

Nevertheless, there were serious flaws and omissions in what was decided at the Congress of Vienna. Its unswerving support of monarchy, rather than republics, flew in the face of the direction of political change, not only in Europe, but also in the Americas and led to considerable pent up republican resentment, especially in France. The interests of ethnic and linguistic groups were almost totally ignored in redrawing the political map, sowing the seeds of much later unhappiness. In Central Europe, the rights of Poles, Czechs, and Slovaks were almost entirely swamped by the demands of the larger powers, whilst in the west, difficulty in bringing Francophone and Germanic peoples together sowed the seeds of continuing conflict. Finally, the most important shortcoming was the failure to engage with the consequences of the terminal decline of the Ottoman Empire in south-eastern Europe. A political vacuum was clearly in the making and, in not addressing the issue, the congress allowed the Balkans and Turkey to become the source of a succession of conflicts that have repeatedly destabilised Europe as whole (Glenny, 1999).

The Monroe Doctrine

In the USA, the Monroe Doctrine was also a response to the new political order after the end of the Napoleonic Wars. The country was constitutionally a republic and working aggressively to reduce the European colonial influence, not only in North America, but in South America as well. The wars at home had diverted the attention of the Europeans, and European resources, away from their overseas interests generally and successive US administrations were determined not to see them re-established anywhere in the Americas. France had already ceded all its holdings to the USA and the last remnants of colonial rule south of the Canadian border in North America were swept away when Florida was purchased from Spain by the USA in 1822. In the same year, the USA formally recognised the fledgling republics in Argentina, Chile, Peru, Columbia, and Mexico.

As a pre-emptive action to dissuade any thoughts amongst the former European colonial powers from trying to reassert their political control, the US President James Monroe with the assistance of his Secretary of State, John Quincy Adams, promulgated to Congress in 1823 what has become known as the Monroe Doctrine. Essentially, it declared that an attempt to interfere with the independence of any state in the Americas would be treated by the USA as a danger to its peace and safety and would be resisted with all necessary force (Poetker, 1967). At the time, it was something of an empty threat, as the USA did not have a sufficiently large and strong navy to back up its warning, but the British were more than happy to lend it substance. Britain was the undisputed naval power in the world in the nineteenth century and was delighted to be able to assist in thwarting any renewed colonial aspirations on the part of France and Spain. For nearly a hundred years it supported the USA and helped make the Monroe Doctrine a military reality until the USA was able to do so on its own (Alvarez, 1924). The thrust of the Monroe Doctrine has proved very enduring. In principle, it is still a key tenet of American foreign policy and one that has long been sustained by the USA, without support from Britain or any other outside power (Scudder, 1939). Somewhat ironically, when the UK used military force against Argentina during the Falklands conflict in 1982, the legacy of the Monroe Doctrine made the USA very hesitant about lending any overt support, preferring rather to be seen as a mediator between the two warring sides.

The Monroe Doctrine is, however, also of much wider importance and strategic significance for the geopolitical balance of power. For the first time, an effective policy was implemented defining a clear-cut theatre of political influence in the world, beyond national boundaries, where a powerful state defended the territory and independence of its weaker neighbours. It has proved a most significant precursor to the present world order, where similar, though more elaborate agreements, such as the North Atlantic Treaty Organisation (NATO) and the South East Asia Treaty Organisation (SEATO), have formed the cornerstone of regional stability in the north Atlantic and south-east Asia respectively, using the USA as a military guarantor.

Geography and the world order

Peace settlements in the wake of major conflicts have always been concluded in the hope of providing for political stability, though as the repeated outbreak of war has shown, success has only ever been very partial. Determination to succeed was never stronger than after the First and Second World Wars. The popularity of the Heartland and other theories developed by geographers in the early part of the twentieth century about the determinants of the balance of political power were eagerly adopted in influential political circles, giving the discipline a unique influence at a key moment in recent history.

Many of the major figures writing about political geography at the time, Mackinder, Bowman, and Curzon, became directly involved in policy-making and brief mention has already been made in the introduction (Chapter 1) of the role of geographers as advisers to the American, French, and UK governments in reconstructing the world map for the Treaty of Versailles (1919), which produced a settlement at the end of the First World War (Figure 9.2 and Box 9.1). Despite some deep scepticism on the part of the USA about the inevitable ascendancy of land power over sea power, Mackinder and his concepts of the Heartland and the geographical pivot of history were important influences on British thinking and made the UK government determined not only to prevent the German Empire from ever again holding Europe and the rest of the world to ransom, but also to see off any other future challenges to Anglo-French hegemony in the continent.

Isaiah Bowman, the Director of the American Geographical Society, was especially influential in determining the future shape of Europe and his ideas are very well known as a result of the his best-selling book, *The New World*, which was first published in 1921 and ran to four editions in the interwar years, selling widely in both Europe and North America. Bowman's view of the world and its post-First World War future was very different from that of Mackinder. He believed in a world order built around a large number of nation states representing distinct ethnic and cultural groupings, their security guaranteed by the League of Nations and, as chief territorial adviser to President Wilson, was in an unrivalled position to press for his ideas to be adopted (Martin, 1980; Smith, 1984 and 2003). It is also interesting, and surely no accident, that one of the major criticisms of the treaties resulting from the Congress of Vienna a century earlier had been the failure to recognise sufficiently the aspirations of minorities of all kinds in Europe.

The decisions made in the Treaty of Versailles, and in a series of subsequent treaties, led to the complete dismemberment of the Austro-Hungarian Empire, as well as to Germany having its eastern and western borders radically trimmed, and to Russia losing control of virtually all the territory it had previously held in Eastern Europe. Out of this radical wing-clipping came a whole host of newly independent states: Finland, Estonia, Latvia, and Lithuania on the eastern seaboard of the Baltic; a much more substantial Poland than in any previous incarnation; and Czechoslovakia, Austria, Hungary, Romania, and Yugoslavia from the now defunct Austro-Hungarian Empire.

Figure 9.2 ***Territorial changes as a result of the Treaty of Versailles, 1919***

Box 9.1

Treaty of Versailles

After the First World War, the Paris Peace Conference was convened on 12 January 1919 until 20 January 1920. The leaders of thirty-two states, representing about 75 per cent of the population of the world, attended, but the negotiations were dominated by the five major powers that had played the major part in defeating the German Empire and its allies. These were the USA, the UK, France, Italy, and Japan. Five treaties eventually emerged from the conference, each named after a Paris suburb: the Treaty of Versailles with Germany, the Treaty of St Germain with Austria, the Treaty of Trianon with Hungary, the Treaty of Neuilly with Bulgaria, and the Treaty of Sèvres with Turkey.

The main points of the Treaty of Versailles were:

- all German colonies to become League of Nations mandates
- the return of Alsace-Lorraine to France
- the ceding of Eupen-Malmedy to Belgium, Memel to Lithuania, and the Hultschin region to Czechoslovakia
- the ceding of the Poznania region, as well as parts of East Prussia and Upper Silesia, to Poland
- the port of Danzig to become a free city
- a plebiscite to be held in northern Schleswig to settle the Danish–German frontier
- occupation by the French and special status for the Saar
- demilitarisation and a 15-year period of occupation for the Rhineland
- Germany to pay £6,600 million in reparations
- a ban on any political union of Germany and Austria
- Germany to accept guilt for causing the war
- the Kaiser and other war leaders to face trial
- the German army to be limited to 100,000 men, with no tanks, heavy artillery, poison gas supplies, aircraft, or airships
- the German navy to have no ships larger than 100,000 tons and no submarines.

Germany signed the treaty, under protest, but the US Congress refused to ratify it. As a result, the Kaiser and other war leaders never stood trial, much to the anger of many in the UK and France.

An important element of the philosophy behind the new states was that they represented broad ethnic and religious groups within the population of Europe, which had previously been denied any real opportunity for self-determination. It was this, rather than any coherent strategy to secure control over Mackinder's Heartland and World Island that dictated policy.

Probably, another underlying reason for the emphasis on self-determination was that this same policy had already been adopted, though with only very moderate success, in the Balkans in response to the collapse of the Ottoman Empire. Undeterred, the surviving victorious powers, minus Russia which had temporarily lost credibility and influence in the wake of the Bolshevik revolution in 1917, decided to press ahead along similar lines in Eastern Europe.

The enthusiasm for national self-determination did not extend much beyond Europe. In Africa and Asia in particular the colonial map was largely reaffirmed, though with European imperial power passing from vanquished to victor. The extensive German overseas empire was effectively shared out amongst Britain and France, though technically they were only acting as guarantors over mandates under the aegis of the League of Nations.

The role of the League of Nations in the Treaty of Versailles settlement is crucial. It started work in 1920 and was conceived as an international body that would arbitrate in future territorial disputes and, if necessary, guarantee the integrity of its member states. The newly formed states, unsurprisingly, left a large number of smaller territorial anomalies unresolved and one of the first tasks of the league was to try to resolve these. It organised plebiscites to determine their future, notably in the Saarland on the border between France and Germany, where the vote went decisively in favour of becoming a part of France. In reality, however, the league was never able to exert the necessary authority to guarantee the peace. The USA refused to join at all, Britain's support was at best equivocal, and only France of the three major victorious powers gave its wholehearted support. The league lurched from one crisis to another, finally losing virtually all its credibility when it failed to force Italy to withdraw from north Africa and its occupation of Abyssinia in 1934.

From the point of view of direct geographical involvement, one of the most significant features of the league's work was the establishment of the so-called Curzon Line, delimiting the eastern border of the enlarged Polish state with Russia (the emergent Soviet Union) (Figure 9.3 and Box 9.2). The line was named after the principle negotiator, Lord Curzon, who had written so extensively about frontiers before the war and was now the British Foreign Secretary (Curzon, 1907). The initial proposal quickly proved unsatisfactory and was revised by Curzon to what is known as the Curzon Line 'B', but this too was highly contentious and

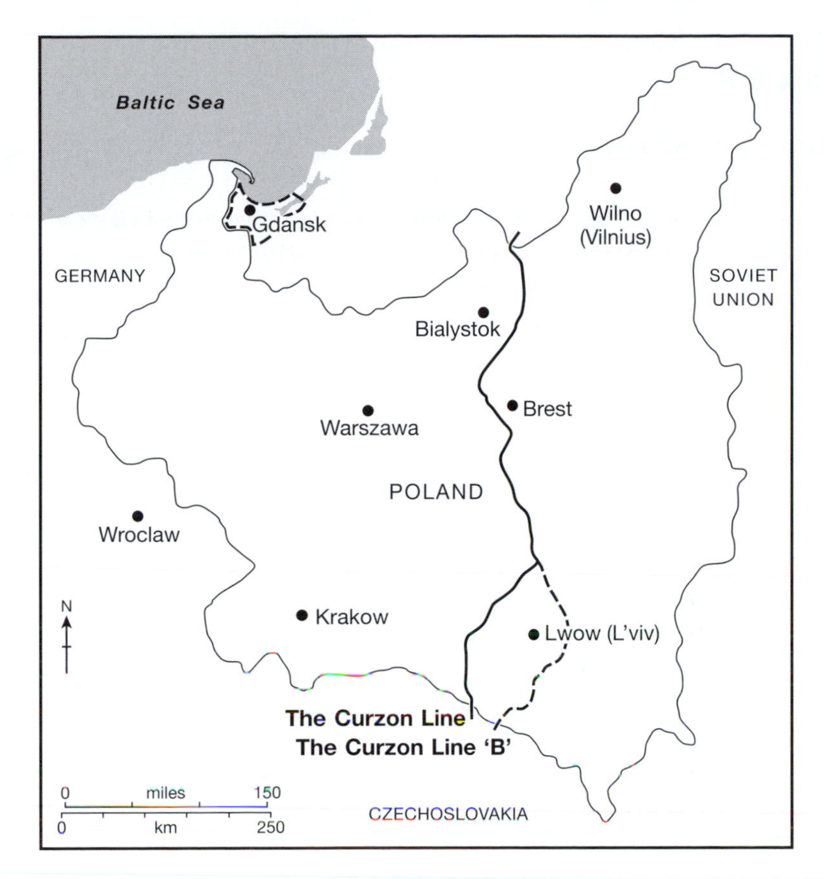

Figure 9.3 ***The Curzon Line defining the border between Poland and the Soviet Union***

led to great tension throughout the interwar period between Poland and the three Allied powers on the one hand, and the recently formed Soviet Union on the other. Ultimately, however, the arguments in favour of the Curzon Line prevailed. It was recognised by all sides as the eastern border of Poland in the settlement in 1945 at the end of the Second World War and still remains so over half a century later.

The Treaty of Versailles was not generally a success and did not endure. As a result, the settlement in Europe at the end of the Second World War (1939–45) was much more pragmatic, reflecting a hard-headed territorial power struggle between the Soviet Union, the USA, and the UK (Blacksell, 1981). The resulting Iron Curtain between the Soviet Union in the east and the other major Allies, the USA, the UK, and France, in the

Box 9.2

The Curzon Line

At the end of the First World War, the Allies agreed that an independent Polish state should be created from territories that had previously been part of the Russian Empire, the Austro-Hungarian Empire, and Germany. The Treaty of Versailles left the question of the final determination of the eastern border of Poland open. As the lands lying between Poland and its eastern neighbours were home to a mixed population of Poles, Lithuanians, Jews, Ukranians, and white Russians, with no single group in a majority, it fell to the British Foreign Secretary, Lord Curzon, on behalf of the Allies, to suggest a border running from Grodno, through Brest-Litovsk, to Lwow, though it was unclear on which side of the border Lwow would be. Subsequently, a second map, known as the Curzon Line 'B', definitely included Lwow in Poland.

The newly constituted Soviet Union refused to accept the Curzon Line and fought a war with Poland between 1919 and 1921, which resulted in a Polish victory and Poland gaining even more territory to the east. However, after the Second World War the Soviet Union annexed all the Polish territory east of the original Curzon Line, which became, and has remained, the eastern border of Poland.

west basically reflected the military situation at the end of the war. The Soviet Union and its client states extended westwards as far as the Baltic in the north and across the Balkans, with the exception of Greece and Turkey, in the south. On either side of the Iron Curtain, two strong military alliances, NATO and the Warsaw Pact, faced each other and maintained an uneasy balance of power for more than four decades.

The Cold War, as the era was generally known, was one of great tension but, in Europe and North America at least, one of relative political stability (Blacksell, 1982; Blacksell and Brown, 1983). As a result, it offered little opportunity for geographers to engage in applied geopolitics. Indeed, partly as a result of what was seen as their malign influence on the development of political strategy in the Third Reich in interwar Germany, and partly as a result of the perceived ineffectiveness – Curzon excepted – in devising any lasting political solutions in the wake of the First World War, geographers working in the field of geopolitics were, to all intents and purposes, sidelined.

Practical geopolitics

Though academic interest in geopolitics faded rapidly in the middle years of the twentieth century, the term began to gain wide general currency for describing the whole process of international, strategic, political decision-making. Its popularity stems in no small measure from the American political scientist, and US Secretary of State during the Presidency of Richard Nixon in the early 1970s, Henry Kissinger. He believed that the USA had traditionally based its foreign policy far too much on moral principle, rather than on achieving a balance of strategic, political power, so that the political and economic might of the USA was always set against a similar force elsewhere in the world (Garrity, 1997). For this reason, he was deeply opposed to the war in Vietnam, and championed strongly the opening up of much closer relations between the USA and China. Once the Cold War had ended, he and other political scientists, in particular Zbignew Brezezinski, argued strongly for re-examining the balance of power within Mackinder's Heartland, in the belief that this would hold the key to the post-Soviet order in Eurasia (Brezezinski, 1997). This geopolitical approach to foreign affairs was highly contentious, even leading to calls for Kissinger to be tried in the International Court of Justice for war crimes, but it undoubtedly extricated the USA from a situation where it was hopelessly overextended militarily in Vietnam and elsewhere in south-east Asia, allowing the country to focus much more clearly on ending the nuclear stand-off with the Soviet Union and bringing the Cold War to a close.

More recently, the term geopolitics has been yet further broadened to cover the search by states to secure access to strategic reserves of mineral resources across the globe. Across the world, in both developed and developing countries, there is an apparently insatiable demand for energy, a demand that in most industrial countries far outruns their available domestic supplies. The USA, despite its huge domestic reserves, has been a net importer of oil for more than half a century and much of its diplomacy in the Middle East and South America has been directed towards ensuring that it is not isolated for political reasons from access to future supplies (Roberts, 2004). It is too simple to argue that this consideration has driven the country to support atrocious tyrannies, such as that of Saddam Hussein in Iraq in the 1980s, but access to oil undoubtedly played a part, as it did when the USA and other Western countries turned on Iraq after it invaded Kuwait in 2000 and, later, when they invaded Iraq for a second time in 2003.

Natural-resource driven geopolitics is also fickle. A change in the balance of energy supplies in the world, as there was in the shift from coal to oil in the middle of the twentieth century, can decisively alter geopolitical calculations. It has been suggested, for instance, that a preference for natural gas over oil, a move that would cause very little technological disruption, would fundamentally alter the balance of power away from the Middle East towards Russia (Doyle, 2004). It is a move that could have considerable attraction for the USA and the other major industrial nations that are finding it increasingly difficult to exert control over political events in the Middle East, West Africa, and other politically unstable parts of the world where major oil fields are located.

Thus, in geopolitical terms, the wheel has in some sense moved full circle. During the age of mercantilism global geopolitics was driven predominantly by the need to secure and protect economic interests overseas. Subsequently, especially in Europe and North America, other considerations, political and military rather than economic, took a more prominent role, but now global economic considerations are once again beginning to exert a decisive influence on the pattern of strategic planning and decision-making.

Key themes and further reading

It is always important to be aware of the impact of ideas on political actions, especially foreign policy. This chapter considers the links between trade and empire, with particular reference to the British East India Company. The birth of practical geopolitics in the modern era is illustrated by the territorial horse-trading at the Congress of Vienna, and by the implementation of the Monroe Doctrine by the USA. The direct impact of Isaiah Bowman, Sir Halford Mackinder, and Lord Curzon on the territorial settlement after the First World War represents a period of unparalleled influence by geographers on foreign policy. Subsequently, the perceived use of the whole concept by geographers for nefarious political ends, particularly in Germany, led to it being consigned to the fringes of political geography until the end of the twentieth century. During this period, however, the term geopolitics and its application to practical politics was commandeered by political scientists generally and it ceased to be any longer in the exclusive preserve of geographers.

A very good introduction to practical geopolitics is *Geopolitics in a Changing World* by Klaus Dodds (2000), while the post-Second World

War reconstruction of Europe and the geography of the Cold War is the main theme of *Post-war Europe: a political geography* by the author (1981). To understand the logic of the wholesale redrawing of the world political map after the end of the First World War, *The New World* by Isaiah Bowman (1928) is indispensable, though Neil Smith (2003) provides an excellent critique of what he terms 'Roosevelt's geographer' and the impact of the underlying philosophy in *American Empire*.

10 Annexing the oceans

The sea! The sea!

(Xenophon, *c*.428/7–*c*.354 BC, *Anabasis* IV.vii.24)

The freedom of the seas

The oceans cover more than 70 per cent of the world's surface and they have been an integral part of the geography of human history since at least the classical era, as Xenophon's enthusiastic acclamation indicates. Yet until the second half of the twentieth century the greater part of them was not subject to any form of political regulation or control. Sailors crossed them at their peril, protected only by force of arms from marauding pirates, or anyone else wanting to interrupt their safe passage (Glassner, 1990). The only semblance of any rules of international behaviour was a de facto acceptance by coastal states, which evolved first in early seventeenth-century Europe, that they had jurisdiction over territorial seas stretching for three nautical miles from their coastlines. The origins of this concept of territorial seas are opaque, but the popular view is that it was driven by defence of the realm and that they represented the extent of the seas that could practically be defended by canon fire from the land.

By the beginning of the twentieth century, the inadequacy of this informal agreement was becoming increasingly apparent and many coastal states were looking at ways of extending their coastal seas. The first to break ranks was the USA, which extended its territorial limit to 12 nautical miles more than a century ago. Other states gradually followed suit, but it was apparent that the main issue was not defence, but rather control over coastal fishing and mineral resources, and that a different kind of regime would be required if this were to be properly addressed.

The focus of interest was the continental shelf (Figure 10.1), the area of relatively shallow seas of very variable extent, stretching out from the coastline. The earliest claim to jurisdiction over the continental shelf was made by Argentina in 1944. The USA was quick to follow suit and, a year later, Harry S. Truman in one of his first acts as president, issued Proclamation 2667, claiming sovereignty over all the seabed and the subsoil of the continental shelf, though not over the waters above and, therefore, fishing rights.

Other coastal states across the world quickly took similar action, but in the absence of any clear definition of what constituted the continental shelf an unacceptable state of confusion ensued. The difficulty was, however, that no international forum existed for arbitrating over disputed claims and, given the nature of sea traffic, bilateral negotiations were a practical impossibility. Eventually, there was broad international agreement that the best way forward would be to call upon the newly formed UN to try to devise a regime for regulating the use and exploitation of the oceans, and to persuade its member states to accept its proposed solution (Couper, 1978).

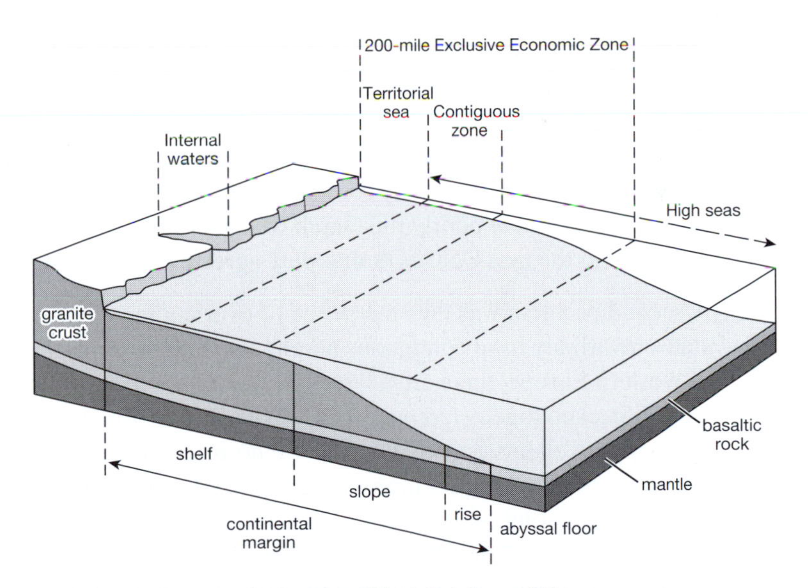

Figure 10.1 ***The physical and political divisions of the oceans***

Source: after A. D. Couper (1978) *Geography and the Law of the Sea*. Macmillan, London, p. 4. Reproduced with permission of A. D. Couper.

UNCLOS – The United Nations Conventions on the Law of the Sea

The first UNCLOS results emerged in 1958 with agreement on the Continental Shelf Convention. This gave states the right to exploit the mineral resources of their coastal waters to a depth of 200 m, or beyond if exploitation was technically feasible. To further such exploitation, states were also to be permitted to build permanent structures, such as oil wells, but with two important provisos. They were not to be accorded the status of islands with territorial waters of their own, nor were they to impose any undue hazard to shipping. The convention also gave coastal states rights to sedentary living species on the seabed, such as shellfish and crustaceans, but not to fish. The absence of any proposals for regulating fishing stemmed from the extensive, and highly convoluted, network of bilateral agreements covering customary fishing rights that already existed between states. It was clear to all sides that any attempt to try to include regulation of fishing in the Continental Shelf Convention would have only served to stymie any agreement.

Once put into practice, the inadequacy of the convention quickly became apparent. It suffered from a number of serious weaknesses, that almost entirely neutralised it as an effective regime. The most fundamental was the failure to define exactly what was meant by the term continental shelf. The convention itself accepted that the 200-m depth limit was entirely arbitrary, by advising that states could claim jurisdiction over greater depths of water if they had access to the technology to do so. It was, therefore, a recipe for instability and constant revision and change, providing endless opportunity for legitimate legal challenge. Indeed, it is a measure of how poorly the march of technology was understood by delegates to the first UNCLOS that they agreed it in the first place.

A second problem was the emergence of overlapping claims. Coastal states invariably have contiguous neighbours and in some areas, such as western Europe, there are a large number of competing jurisdictions closely packed together, rendering the 200-m outer limit meaningless. Overlapping claims quickly became the norm, rather than the exception, offering yet another opportunity for almost unlimited litigation. The most obvious solution initially appeared to be a median line between two claims, but the problem was actually chronic, because there was no agreed mechanism for defining the coastal baseline from which such median lines should be measured. Coastlines are never straight and, depending on which particular point was chosen, quite different median lines and definitions of national continental shelves would be produced. Furthermore, there was no agreement about from where baselines should be measured. Inland waters, such as Hudson's Bay, are normally considered as part of the land area, but the Gulf of Bothnia separating Sweden from Finland, Estonia, and Latvia in the Baltic Sea, is not, showing that the basis for defining an inland water is far from consistent. Countries like Norway, which have deeply indented coastlines, usually measure their baseline from a line joining the major promontories. Island states, such as Indonesia and the Philippines, pose especially difficult choices, but their baselines are now nearly always taken as straight lines encompassing the whole of the archipelago. The detail of the decisions about baselines are often economically vitally important, deciding which state has jurisdiction over valuable natural resources. Modern technology in the form of accurate positioning, satellite imagery, and GIS has the potential to solve some of these problems, though in practice it serves only to emphasise the shortcomings of the available charts (Cleverly, 2004).

The way in which the map of a coastal state can change if internal waters are taken as being part of the baseline and the territorial waters are included are clearly illustrated in Figure 10.2. The most obvious additions are all the waters around the Outer Hebrides and the Western Isles, which, with the addition of the territorial seas, mean that Northern Ireland is no longer separated politically from the rest of the UK.

Islands also pose special problems, since not only can they have valuable territorial waters associated with them in their own right, they can also form a point on a national baseline, sometimes increasing substantially the extent of the territorial waters of a coastal state. The continuing dispute over the islet of Rockall, which has been claimed inconclusively by the UK, Ireland, Iceland, and Denmark is evidence of the lengths that states can go in order to substantiate a claim to even the most inhospitable and, apparently, worthless ocean landfalls (Box 10.1).

A second UNCLOS was held in 1960 and only narrowly failed to reach agreement on defining more unequivocally an outer limit for territorial seas, though the net effect was to leave the whole issue unresolved and little short of anarchy as far as fishing and mineral rights over the continental shelf and the rest of the high seas were concerned. Increasingly, states began to act unilaterally, generating increasing levels of international tension. The most notorious of these conflicts were the three Cod Wars between the UK and Iceland between 1958 and 1976.

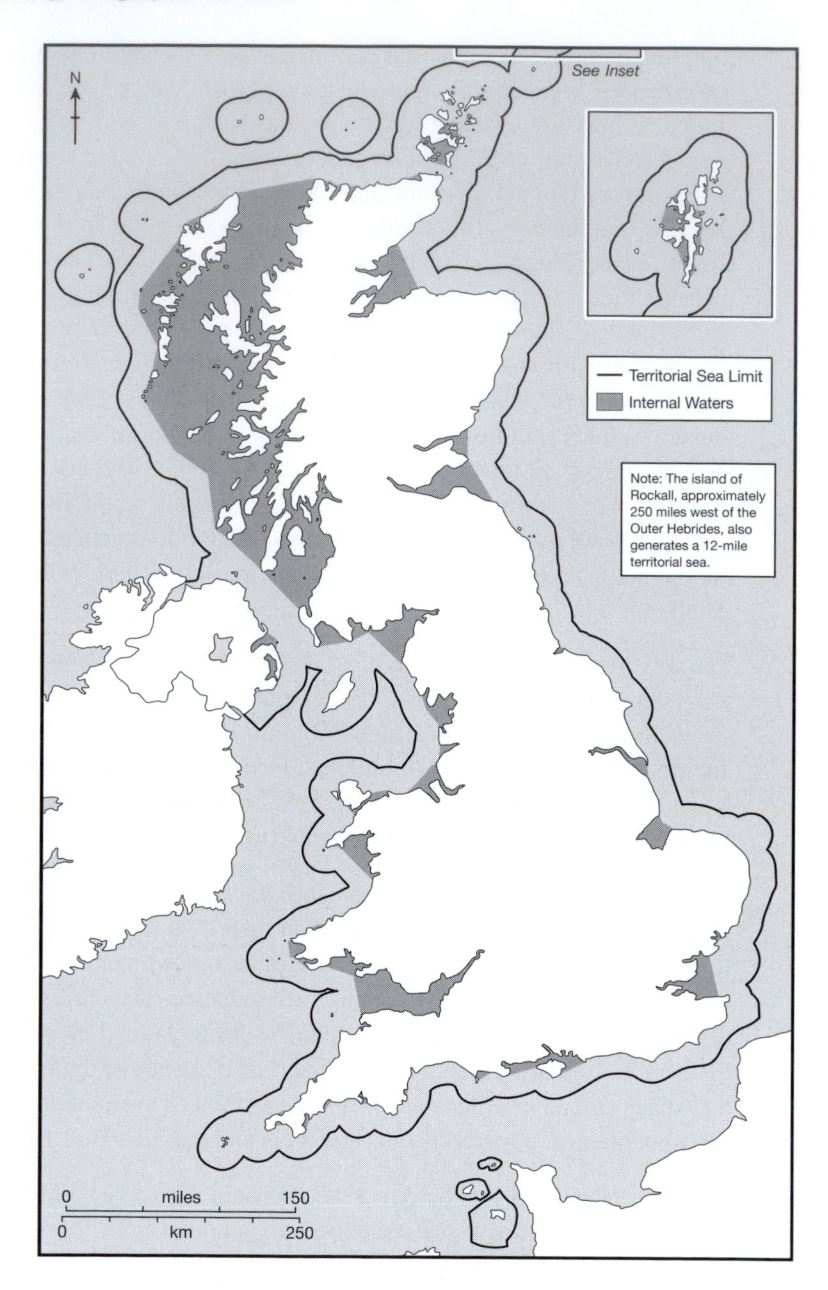

Figure 10.2 ***UK showing internal waters and territorial seas***

Source: after UK Hydrographic Office (2004). Reproduced by permission of the Controller of Her Majesty's Stationery Office and the UK Hydrographic Office (www.ukho.gov.uk).

Box 10.1

Rockall

Rockall is an isolated, uninhabited, pudding-shaped islet in the middle of the North Atlantic Ocean. It is only 19 m high, 25 m across, and 30 m wide and is located 57° N and 13° W, roughly 300 miles from the coasts of the UK, Ireland, and Iceland. The UK, Ireland, and Denmark, by dint of it once having governed Greenland as a colony, have been arguing for many years over whether the islet is large enough to justify having territorial rights and, more importantly, which country has the rights to them. At issue is not really the sovereignty of Rockall itself, but its territorial waters and the fishing grounds that go with them and any oil, gas, and minerals that can be extracted from the seabed.

The British originally claimed Rockall in 1955 and a number of so-called annexation forces climbed the rock to legitimise the claim. In 1971, the UK government passed an Act of Parliament incorporating Rockall into Invernesshire. To reinforce the claim, they installed a navigational beacon and banned all craft from approaching within 50 miles. In the 1980s, on two separate occasions, a British soldier climbed onto the rock and remained for a month, underlining the government view that it was an island and not uninhabitable. Not to be outdone, the environmental pressure group, Greenpeace, landed a party on Rockall in 1997, which stayed for 42 days, replaced the British navigational beacon with a solar-powered one, and claimed that they had saved the rock from oil development and industrialisation. There is still no international agreement on the status of Rockall.

The first broke out in 1958 when Iceland increased the extent of its territorial waters from 3 to 12 nautical miles. Deprived of access to port facilities on the Icelandic mainland, the British deep-sea fishing fleet was forced to accept the extension, but did so only after some noisy sabre-rattling. The second conflict in 1972–3 was altogether more serious. Iceland extended its territorial claim from 12 to 50 nautical miles, effectively excluding the British fleet entirely from its main deep-sea fishing ground for cod. Once again, after a tense stand-off, a compromise was agreed. For a two-year period, the British fleet was allowed to take 130,000 tonnes of fish from the newly defined Icelandic waters. However, when this agreement ran out in 1975, an emboldened Icelandic government extended the territorial waters to 200 nautical miles, with no access for British fishermen. A compromise proved impossible; gunboats squared up to each other, and a number of warning shots were fired. After a winter of highly dangerous brinkmanship, the UK had to accept that it could do little to reverse the Icelandic action and reluctantly withdrew, tacitly accepting the new limits. Of themselves, the Cod Wars were a source of considerable international hilarity for all but the two countries directly involved, but at a more fundamental level they emphasised all too clearly the urgent necessity for international agreement.

In fact, even before the two conflicts in the 1970s, the urgency of the situation had been recognised by the UN and UNCLOS III was convened in 1973. The conference met every year from 1973–82, usually twice a year, and eventually agreed a draft convention at Montego Bay in Jamaica, covering all the major issues associated with an effective political regime for the oceans (Juda, 1996). The convention covered six broad areas: navigational issues, the exploitation of natural resources, deep-seabed mining on the high seas, protection of the marine environment, marine scientific research, and the settlement of disputes.

Navigation

It was crucial that extent of national territorial seas be regularised, and the convention proposed that there should normally be a 12-nautical mile limit to territorial seas, replacing the traditional 3-mile limit and other unilateral claims of up to 200 miles made by Ecuador and some other states in Latin America. Within this zone state laws apply, but with the important exception that all ships enjoy the right of innocent passage, which means that they should not prejudice the peace, good order, or the security of the coastal state in question. Beyond the 12-mile limit, states are also free to claim a contiguous zone, up to a total of 24 nautical miles, for customs, fiscal, immigration, and sanitary purposes, thus enabling the authorities to pre-empt illegal access into their territorial waters.

Extending territorial waters in this way, of course, raised new problems, the most critical being the status of ocean straits. The convention defined more than 100 straits used for international navigation, such as the Strait of Gibraltar, linking the Mediterranean to the Atlantic Ocean between Morocco and Spain, and articulated a right of international transit passage through them, including submerged submarines!

The convention also defined a new concept in international law, the archipelagic state, to allow mid-ocean archipelagos, such as the Maldives in the Indian Ocean, to enjoy similar territorial rights to other island and coastal states. For the first time, they had the right to draw archipelagic baselines connecting the outermost islands, though subject to stringent limits. The overall length of the baseline was normally not to exceed 100 nautical miles, though exceptionally this could be extended to 125 nautical miles, and the upper and lower ratios of water to land area within the baseline had to be in the ranges 1:1 and 9:1 respectively. The reason for these limits is to prevent some island groups, such as those in the Bay of Bengal, declaring themselves archipelagic states and, thus, extending the area covered as of right by their territorial seas. Not only would such a move restrict access to fishing grounds and other resources, it could also impair access to the high seas for other states.

All other parts of the oceans are defined by the convention as the high seas. Here, states have complete freedom of navigation and the right to over fly, as well as to lay pipelines and submarine cables, construct artificial islands, fish, and undertake scientific research. It is a recognition that, even if some form of regulation were to be desirable, it would be virtually impossible to enforce, as there is no international body with the necessary legitimacy or resources. It also acknowledges that the high seas are a global resource and that all nations have a right to benefit from them.

Exploiting natural resources

An end to the uncertainty created by the first UNCLOS Convention in 1958 about the right of states to exploit the resources of their adjacent continental shelf was a priority for the third conference. As a solution, it proposed that all coastal states should be able to claim the exclusive right to exploit the mineral and energy resources of an area extending 200 nautical miles from their baselines, to be known as the Exclusive Economic Zone (EEZ). In all other respects the waters of the EEZ remained part of the high seas, with all its freedoms of navigation and the right to conduct military activities on, over, and under the surface of the water. Almost immediately after the convention was published in 1982, all coastal states moved to take advantage of the proposal, in many cases giving private companies the confidence for the first time to invest in extremely expensive exploration, mainly for oil and natural gas. States also have the right to fisheries in their EEZ, though with special restrictions: for highly migratory species, such as tuna; for marine mammals; and for species of fish like salmon and eels, which spend part of their life in freshwater rivers.

On their own, the EEZs did not solve all the problems of jurisdiction over the exploitation of natural resources on the continental shelves. In some cases, the continental shelf clearly extends for more than 200 nautical miles, even though a precise boundary is always very difficult to define. Where this occurs, the 200-nautical mile limit can be extended, but permission is not automatic, being decided upon by the UN Commission on the Limits of the Continental Shelf and usually involving some contribution by the benefiting state to the common good of all nations. There is also a continuing problem of overlap. There can be substantial differences in the area claimed as an EEZ, depending on which point on the baseline is taken as the reference point. The uncertainty frequently necessitates bilateral negotiations between neighbouring states to agree the boundary between their EEZs, no easy task, as the protracted arguments between the UK and Ireland, and the UK and France have clearly demonstrated.

Islands, so long as they are naturally formed areas of land, above water at high tide, are entitled to their own territorial waters. As can be seen from Figure 10.2, these island waters can add significantly to the area of sea over which a state has exclusive jurisdiction.

Deep-seabed mining on the high seas

The most contentious of all the areas in the convention was the regime for mining the deep seabed, beyond the EEZ. The original proposal in the convention was for an international regulatory regime under the UN International Seabed Authority. It would control the levels of production and provide for a mandatory transfer of the mining technology to developing countries, as well as a system of compensation for those developing countries whose land-based production of minerals, such as copper and nickel, was likely to be adversely affected by deep-sea mining.

In the early 1980s, there was considerable excitement about the possibilities for deep-sea mining, mainly from deposits of metallic nodules on the ocean floor, containing nickel, cobalt, copper, manganese, and other metals. Subsequently, this interest has waned somewhat, as estimates of the extent of the deposits have declined, and the costs of recovery have risen.

All the major industrial countries immediately objected strongly to the proposals and refused to countenance signing up to the convention as it stood. Essentially, they objected to the whole concept of there being a UN International Seabed Authority, which would have the power to restrict their freedom to prospect and mine on the high seas, and to require them to provide for technology transfer. Negotiations dragged on for a decade, but in 1994 agreement was reached and all the industrial nations, with the exception of the USA, signed. The USA, the largest industrial economy in the world, is still not prepared to countenance any restrictions and, therefore, refuses to sign up to the convention as a whole. At the moment, the issue of deep-sea mining is of little immediate importance, because there are no serious commercial operations, but in the future this is bound to change. One has only to look at the way in which the possibilities of exploiting shallow coastal waters were being dismissed as recently as the middle of the twentieth century, to see how quickly developments in technology can bring about change.

Protecting the marine environment

The marine environment, in particular the high seas, is obviously highly at risk from pollution and the UNCLOS Convention requires all states to adhere to basic standards, usually those already laid down by other international organisations, such as the International Maritime Organisation. The states themselves are responsible for policing and enforcement within the area covered by their EEZs, while the UN International Seabed Authority is responsible for the situation on the high seas. As always, the most difficult issue is the whole question of enforcement, though as technology improves states are becoming increasingly adept at tracking down offenders, especially ships that dump oil into the open sea.

Scientific research

The convention pays considerable attention to marine scientific research, emphasising its importance for the future sustainable exploitation of the oceans, but without defining precisely what it should encompass. Responsibility for monitoring and managing research within EEZs rests with individual coastal states, but with the International Seabed Authority in the high seas. However, all governmental authorities are committed not to withhold without very good reason requests to undertake research activities, making the maritime position much more open, and very different, from that on land.

Settling disputes

Disputes arising from the interpretation and application of the UNCLOS Convention are a matter for either the UN Law of the Sea Tribunal based in Hamburg, the International Court of Justice in the Hague, or specially convened arbitration tribunals. States have considerable freedom to choose which route they wish to use and also have the option to exclude altogether certain types of dispute, such as those involving delimitation, military activities, and those arising from the UN Security Council exercising its legitimate functions. In short, the regulatory regime is far from being a comprehensive one and its effectiveness has yet to be fully tested in practice.

The EU Common Fisheries Policy

As far as regulating fishing was concerned, the UNCLOS agreements were never likely to prove helpful when it came to managing fisheries in parts of the world, like Western Europe, where there are a large number of small coastal states, all with important fishing industries. Recognising this, the Common Fisheries Policy (CFP) of the EU was first agreed in principle in 1970 and proposed giving all member states unrestricted access to the territorial waters of each other until more detailed arrangements for fishing were agreed. At the time the EU comprised just six members, with little overlap between areas covered by their respective fishing fleets, but in 1970 the impending accession of Denmark, Ireland, and the UK meant that not only would the overall size of the fishing industry rise sharply, but there would also be substantial potential conflict, if there was unrestricted access for all member states in every part of their combined EEZs (Wise, 1984).

From the outset, it had been decided that a comprehensive policy must be introduced by 1983 and, for a decade, there were increasingly fraught negotiations trying to reach agreement. When the deadline arrived, the member states had still not succeeded, but the reality of there being no national or EU-wide regulation quickly concentrated minds and the policy was agreed soon afterwards. In broad terms, the policy allowed member states to have exclusive control over fishing within the 12-nautical mile limit of their territorial seas, but allowed unrestricted access within the rest of their EEZs, subject to policies on the size of catches and the number of boats in each national fishing fleet, all to be determined centrally by the European Commission.

From the outset, the CFP was a source of friction between the member states themselves and with the European Commission and, predictably, the disagreements were greatest amongst those with the largest fishing industries, Denmark, France, Ireland, and the UK. Unfortunately, agreement became progressively more difficult as time progressed, especially once Spain and Portugal, both countries with large fishing fleets, joined the EU in 1986. Nevertheless, the EU has persisted with the policy and gradually developed a system of agreeing annually total allowable catches (TACs) for all the major commercial fish species, including quotas for specific fisheries, such as the North Sea, the Irish Sea, and the Southwest Approaches. It has also taken parallel actions to reduce the size of the fishing fleet, though there are still over 90,000 registered fishing vessels of varying sizes operating officially in EU waters.

The main challenge for the CFP is not the size of the market for fresh fish. Demand is far larger than can by caught in EU waters and the EU imports over 4 million tonnes of fresh fish annually, more than a third of its needs. Unfortunately, catches at present levels are unsustainable, so that the main goals of the CFP are determining and enforcing catch quotas which will ensure that there is a viable fishing industry in the EU in the long term. To this end, it has also invested heavily in promoting aquaculture and fish farming, but the scale of these operations nowhere near compensates for the reductions it needs to impose on catches in the open seas.

Managing decline is always a very difficult exercise and in the context of the CFP it is doubly so, because of the tensions between member states. Countries with large fleets, such as the UK and Ireland, resent the fact that many of their local fishing grounds are now open to boats from other EU states, which they see as undermining their domestic fishing industries. It is a situation that has been further hugely exacerbated by the worldwide changes stemming from the UNCLOS Convention. The establishment of the 200-nautical mile EEZs as the norm for territorial waters has meant that fishermen everywhere have been excluded from most of their traditional fishing grounds, forcing them to concentrate their efforts closer to home. However, in areas like Western Europe, the local seas simply do not support the fish stocks to underpin the size of industry to which the fishermen had become accustomed.

Conversely, of course, in many other parts of the world, the fishing industry has been able to develop locally in a way that was impossible previously. Many developing countries in Africa, Asia, and South America now have much better technology available to them and are able to enjoy access to fish stocks that were previously a global resource, open to all.

It is important to remember that, despite the political revolution witnessed in the coastal waters of the continental shelves, the bulk of the world's oceans are still designated as the high seas with no restrictions on their exploitation for fishing. However, the fish species living here are different from the shallow coastal water species, like cod and haddock, that have traditionally been the staple of the industry and the deep water species living in the oceans of the high seas require substantially different technologies in order to catch them. They are also unfamiliar to consumers and, therefore, do not necessarily have the immediate appeal of shallow water species. Furthermore, as the population of the world inexorably increases, and with it the demand for fish, the pressure on stocks in the high seas is also going to steadily increase and this will, ultimately, force restrictive management regimes to be introduced here as well.

The amazing fact about the political annexation of the oceans that has happened so rapidly since the middle of the twentieth century is that it was so long in coming. Throughout all the economic, social, and political upheavals of the industrial revolution the management of the oceans remained essentially unchanged; the changes of the last fifty years have essentially been a catching-up exercise beginning to bring the world's oceans face to face with the realities of the modern world.

Key themes and further reading

The systematic incorporation of the oceans into a formal political framework since the middle of the twentieth century has been one of the most important recent changes to the world map. The way in which the ocean space has become progressively differentiated into a series of distinct zones is an important topic for political geography. The zoning closely mirrors the capacity of technology to exploit marine resources, running from a baseline, distinguishing the open sea from inland waters, to territorial waters, the contiguous zone, the EEZ, and the high seas. The role of the UN in overseeing the process of differentiation through a series of international treaties, culminating in UNCLOS III, has been highly innovative and significant, not least in clarifying the legal status of islands within the new maritime regime. Elsewhere, groups of states have concluded binding treaties, mostly governing fishing rights. The most ambitious of these is the Common Fisheries Policy of the EU, which has done much to draw attention to both the strengths and the weaknesses of attempts at international management of the oceans.

The most readable political geography of the oceans is M. I. Glassner's (1990) *Neptune's Domain: a political geography of the sea*. For those wanting a more detailed and formal legal account, *International Law and Ocean Use Management: the evolution of ocean governance* by L. Juda (1996) provides all the information one is likely to need in a most authoritative text. The tortuous history of how the EU Common Fisheries Policy was agreed is described by Mark Wise (1984) in *The Common Fisheries Policy of the European Community*.

SECTION C

Beyond the state

11 Globalisation and the theory of world systems

> Young man, there is America – which at this day serves for little more than to amuse you with stories of savage men, and uncouth manners; yet shall, before you taste of death, show itself equal to the whole of that commerce which now attracts the envy of the world.
>
> (Edmund Burke, *Speech on Conciliation with America*, 22 March 1775)

Globalisation

Political and economic interconnectedness in the world is nothing new, but in the twenty-first century it has reached new heights, reflecting the unprecedented technological advances in recent years. New developments in information technology and transport, allied to cheap and abundant sources of energy, have, for practical purposes, made the world a smaller place and forced societies at all levels to reassess their images of themselves and how they function (Harvey, 1989). The time–space compression, or the reduction in the barriers of physical distance by the introduction of ever faster means of communication and travel, has led to what Thrift (1995) has described as a hyperactive world, where the sheer volume and speed of transactions across the globe, and across space, has created a totally new political and economic landscape.

The revolution, which Edmund Burke foresaw over two centuries ago, is frequently, and often somewhat loosely, referred to as globalisation, though it is far from being a single, simple process. It is, rather, the convergence of a number of varied and quite disparate changes (Waters, 1995). These changes have necessitated a radical reappraisal of political geography and, in this context, there have been calls for a completely new approach to geopolitics, reasserting the crucial symbiosis between politics and economics, each of which is a necessary prerequisite for the successful application of the other in international policy-making (Agnew and Corbridge, 1995).

At times, the concept of globalisation has led to somewhat extravagant claims being made about the scale and novelty of the revolution that is in train, not to mention its likely impact. While it is true that the role and power of the nation state have begun to change, predictions of its imminent demise in the face of a challenge from global, transnational corporations are decidedly premature and national forces clearly still remain extremely important and influential (Hirst and Thompson, 1996). At issue is a debate about the precise nature of the processes at work and the extent to which the world is becoming more internationalised, or more globalised (Dicken, 1998).

Internationalisation involves no more than the spread of economic activities across national boundaries and is, essentially, a quantitative process, leading to a more extensive global pattern of economic and commercial activity. Globalisation, on the other hand, is a more fundamental, qualitative change, producing novel patterns and processes of production and exchange and leading to a change in the whole structure of the economic landscape (Hodder, 1997). In reality, of course, such a rigid distinction is false in that both processes coexist, side by side, each to some extent a product of the other. The internationalisation of economic activity has encouraged novel solutions in both production and marketing, which have transcended national political boundaries and made globalisation a more distinct reality. Even so, the distinction between the two concepts is important, because both are highly uneven across time and space, with their absolute and relative distributions in a constant state of flux. Changes in one part of the world are rapidly diffused across the globe, underlining the interdependency of the whole economic system.

Nothing illustrates the scale and impact of the changes better than the progressive deregulation of global money markets since the end of the Second World War. Previously, world trade had been hidebound and very hampered by a multitude of national currency regulations, but following the UN-brokered Bretton Woods agreement in 1944, international currency convertibility gradually became the norm. First of all, as part of the US-led post-war economic reconstruction, the Organisation for European Economic Cooperation (OEEC) established fixed exchange rates for Western European currencies against the US dollar, thus allowing Western European countries to trade freely with North America, and with each other (Blacksell, 1981). Later, the International Monetary Fund (IMF) was specifically charged with providing international support for weaker economies and currencies, underpinning the emergent, new financial order and extending the possibility of less restricted trade to other parts of the world.

The OEEC was extremely successful, but limited in its geographical scope and it was succeeded in 1961 by the Organisation for Economic Cooperation and Development (OECD), extending membership to most of the larger trading economies in the non-Communist world (Box 11.1). The success continued and, by the late 1960s, it was becoming clear that most of the major Western national economies had become strong enough economically to fend for themselves and that fixed exchange rates against the US dollar were an unnecessary anachronism. In addition, the scale of world trade and the relatively weak state of the US economy in the early 1970s meant that the USA was no longer in a position to allow the US dollar to be used as a universal reserve currency. Most currencies in the Western world were, therefore, allowed to float

Box 11.1

Organisation for Economic Cooperation and Development (OECD)

The OECD grew out of the Organisation for European Economic Cooperation (OEEC), which was formed to administer US and Canadian aid under the Marshall Plan for the reconstruction of Europe after the Second World War. Since it took over from the OEEC in 1961, the OECD's vocation has been to build strong economies in its member countries, improve efficiency, hone market systems, expand free trade, and contribute to development in industrial as well as developing countries.

The founding and early members of the OECD were all countries in Western Europe and North America: Austria, Belgium, Canada, Denmark, France, Germany, Greece, Iceland, Ireland, Italy, Luxembourg, the Netherlands, Norway, Portugal, Spain, Sweden, Switzerland, Turkey, the UK, and the USA. Membership has subsequently spread much more widely across the trading nations of the world and now includes: Japan, Australia, New Zealand, Finland, Mexico, South Korea, as well as four former Communist states in Europe: the Czech Republic, Hungary, Poland, and the Slovak Republic.

freely against each other and to find their own relative values. The world financial system remained robust in the face of the change and, as a result, governments were encouraged to relax still further national controls on the free movement of currencies, to a point in the early 1980s where currency controls virtually disappeared entirely throughout the Western world.

The removal of restrictions on the movement of money transformed financial institutions. No longer necessarily under the dictatorship of national governments, they were free to locate and trade as they wished and a competitive global financial market rapidly began to take shape (Leyshon and Thrift, 1997). Money can now be moved around the world almost without any restriction, so long as the process does not infringe the criminal laws of the countries concerned. In a sense, a market has been encouraged to develop between states and other political jurisdictions, with financial institutions, such as banks and investment companies, competing to find the locations with the least punitive fiscal regimes. There is now a host of micro states, many of them former British and French island colonial territories, which have developed as important financial centres by acting as tax havens, where individuals and companies can avoid paying tax in the major industrial countries where most of them do business and are located (Figure 11.1). Tax havens are now to be found in every part of the world, so that most countries have an easily accessible place where money can be deposited to avoid paying tax. There is also a growing number of larger states that are seeking to emulate the notorious secrecy of Switzerland, which has acted as a no questions asked and no tax levied bolt hole for money from all parts of the world for more than a century. Not only does this secrecy mean that Swiss banks act as an impenetrable front for often ill-gotten gains, it also unfairly penalises citizens of some of the poorest countries in the world by depriving them of resources that are rightly theirs and compounding their poverty.

Transnational corporations (TNCs)

Transnational corporations with their operations based in a number of different countries across the world have been the business response to the greater financial freedom that the world economy now enjoys (Coe *et al.*, 2004). They dominate world trade, with over 50 per cent of the total volume of trade of the USA and Japan being accounted for by

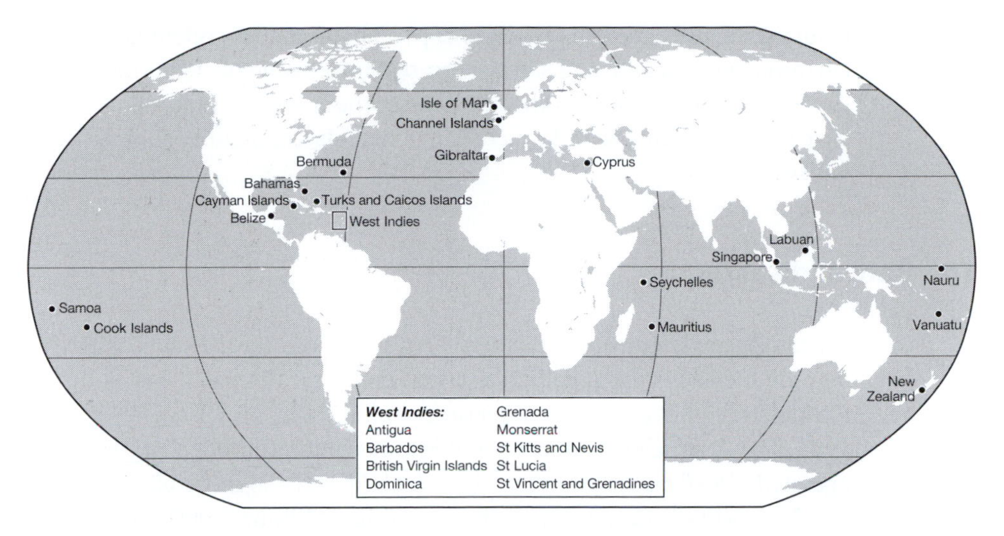

Figure 11.1 ***Tax havens across the world: small states where money can be deposited to avoid national taxation elsewhere***

the international movement of goods and services within their largest companies. Some idea of the size of TNCs, and their smaller cousins, the MNCs (multinational corporations, with operations in more than two different countries), is evident from the fact that the largest of them, corporations such as Ford, Exxon, Mitsubishi, and International Business Machines (IBM), have turnovers greater than those of countries like South Africa, Greece, or Portugal (Knox and Agnew, 1998). Their sheer size makes them formidable players on the global political stage in their own right. They are also overwhelmingly based in the traditional seats of economic power and wealth, with about 90 per cent of TNC core operations located in the USA, the EU, and Japan, though MNCs in a number of other countries, like South Korea, are beginning to challenge the supremacy of the traditional triad.

The somewhat contradictory national concentration of the economic power of TNCs within the traditional industrial heartlands gives some clue as to how they are structured. Although TNCs are certainly global, for a company such as IBM, which has production sites in 84 different countries worldwide, the profits from these operations flow very strongly back to the USA, where the original hub of its activities, and the bulk of its shareholders, are still located. The attraction to IBM of its far-flung empire is partly that it makes the corporation better placed to exploit new market opportunities, but also partly that it gives it more options for reducing costs by capitalising on cheaper labour and less onerous environmental, health, and safety requirements. Such considerations can be very significant if companies and corporations at all levels have the opportunity to reduce costs in a fiercely competitive global market.

The Bhopal disaster

An extreme example of the dangers TNCs can pose for the states where they are located is the appalling disaster visited on Bhopal, a city with a population of over 1 million in north-central India, on 3 December 1984. The American chemical company, the Union Carbide Corporation, operated a pesticide plant in the city which leaked a highly toxic cloud of methyl isocyanate into the atmosphere, killing 2,000 people immediately and injuring at least 600,000, of whom more than 6,000 have subsequently died. The tragedy was made particularly devastating because the leak went undetected for at least an hour, and because neither the local population nor the local health officials had been given any training in how to respond to such a disaster and were, therefore, unable to apply the basic, and very straightforward, precautions that would have neutralised the worst toxic effects of the gas.

Union Carbide Corporation's main defence against charges of criminal negligence was that it was not actually directly responsible, as the plant was operated by a local Indian company, Union Carbide India Ltd, which had built the plant and was wholly in charge of health and safety. What the Union Carbide Corporation failed to mention was that it was the majority shareholder in the Indian company and, thus, in an unassailable position to ensure that proper operating standards were in place.

The Indian government successfully sued the corporation for $470 million in compensation, but the amounts paid to the victims were pitifully small: $1,300 for a death and $550 for those injured. What is more, corruption within the Indian government has meant that nearly half of the settlement has yet to be distributed. The plant was shut down immediately after the accident, but neither Union Carbide India Ltd, nor the Union Carbide Corporation, has managed to complete the clean-up operation and a cocktail of toxic chemicals is still leaking into the local environment, posing yet a further threat to the already beleaguered population of Bhopal.

Similar, though thankfully less disastrous, incidents have occurred across the world, most of them in developing countries. They are undoubtedly in large part an unsavoury consequence of the way in which many TNCs manage their global operations, even though it would be wrong to read into this that there have been no such incidents in the USA, the EU, or Japan. Such incidents have occurred there, but they have tended to be on a smaller scale and less serious. The one real exception to the general rule, however, is the former Soviet Union, which before its demise perpetrated massive environmental destruction in the name of industrial development, especially in the more remote of its constituent republics in Siberia (Saiko, 2001).

Oil exploration

Worldwide oil exploration represents the political dilemmas faced by those TNCs whose main business is to find and develop natural resources. On the one hand, few developed countries in the world have sufficient reserves to satisfy the needs of their major domestic oil companies, forcing them to seek opportunities abroad. On the other, these same companies have invaluable technical and business expertise to offer less developed countries to help them develop and realise the economic value of their oil and other natural resources. Nevertheless, most of the major oil corporations have come into serious conflict with governments at some time or other in their pursuit of new reserves, none more so in recent times than Royal Dutch Shell in Nigeria.

Shell produces nearly half of Nigeria's oil and in August 2004 was pumping out about 1 million barrels of oil a day in the country. It is the largest single contributor to Nigeria's exports and, as such, has to have a close working relationship with the central government. The dilemma is that that the central government's own legitimacy and hold on power is tenuous, as is often the case in the developing world, and this can throw an oil exploration company, like Shell, into conflict with dissident, irredentist groups, fighting to assert their autonomy. It was in the face of just such a threat that, in 1993, Shell had to suspend its operations in the Ogoni area of the Niger Delta, which is where most of its drilling operations are located. The company was accused by the local people of conniving with the central government to destroy their land and their way of life in its drive to exploit the rich oil and natural gas reserves. Shell's defence was that it was investing in the future of the local people by bringing wealth and development to the region. The corporation was caught in the middle of what was a virtual civil war in the delta region and, although the immediate danger of war breaking out has now receded somewhat, the future of oil exploration is still hotly contested and Shell is still seen by many people locally as little better than the agent of a repressive central state.

An interesting consequence for geography of the position Shell has found itself in, stemmed from the fact that the company was and remains a major sponsor of the Royal Geographical Society in the UK. Many geographers at the time felt that the RGS should sever all links with the company in protest at its activities in the Ogoni region and the issue was the subject of a heated debate at the society's annual general meeting in 1994. In the event, the membership decided to continue accepting support from Shell, swayed by the argument that the social and economic benefits that the company brought to the region outweighed the tacit support that its presence in Nigeria gave to the repressive regime. The dilemma faced by both the company and the RGS provides a classic example of the difficulties caused by the fact that as economic systems become more globalised, political institutions often struggle in their wake.

Developmentalism and development

What is called 'the error of developmentalism' is a phrase first coined in 1974 by Immanuel Wallerstein in his monumental Marxist analysis of the evolution of the world economy (Wallerstein, 1974). It refutes the liberal notion that states develop through a series of discrete stages, from traditional to complex societies, which was articulated most tellingly by the American economist Walter Rostow (1971) and widely accepted in the Western world at the time as the orthodox interpretation of the development process. Wallerstein argues that the evidence for such an automatic progression in the development process simply does not exist. Rather the reality for most states in the developing world is that they are stuck in an unequal exploitative relationship with the states in the developed, industrialised world and that the relative economic positions are unlikely ever to change (Dos Santos, 1973).

Indeed, the whole concept of development, with its intrinsic promise of future wealth and prosperity, has been widely criticised as inherently fraudulent, preserving the essentials of European colonial exploitation in a less obvious form, as overt colonialism fell out of political favour in the second half of the twentieth century (Escobar, 1995). Initiatives, such as the 1940 Colonial Development and Welfare Act in the UK and the 1946 Investment Fund for Economic and Social Development in France were as much agents for preserving the economic status quo as they were attempts to define a new political relationship between the UK and France and their hitherto dependent territories (Watts, 2000).

The persistence of the inequalities inherent in the relationship between the developed and the less developed worlds can still be seen in the difficulties encountered by the EU in developing an acceptable and equitable relationship with the former colonial territories of its member states. Since 1964, there have been a series of conventions signed between the EU and over fifty former dependent territories of its member states, mainly in Africa, the Caribbean, and the south Pacific, the most recent being the Fourth Lomé Convention, signed in 1989, which has subsequently been revised and updated.

All the conventions were, broadly, reciprocal agreements giving unrestricted access for exports from the former dependent territories to the EU, and also unrestricted access to markets in the former dependent territories to the EU. The agreements have proved very advantageous to the EU, because of the guaranteed access they gave to the EU to minerals and other raw materials in a large part of the developing world. It quickly became clear, however, that too little was being done to protect the price of exports from the former dependent territories and a series of STABEX agreements, now covering 50 different products, have since been agreed to stabilise the level of their export earnings and give their primary industries a more equitable return on what they produce. A separate agreement, SISMIN, has been agreed to cover the price of mineral exports. In spite of these agreements, there has still been a considerable amount of criticism of the EU for exploiting its relationship with the former dependent territories, though a good proportion probably stems from the jealousy of other industrial countries that are not part of the agreement (Blacksell, 1981).

In the eyes of many commentators, much of what has been represented as development is no more than a cynical ploy to preserve and perpetuate economic privilege. Haraway (1991) argues that the concept is largely constructed through keywords, what she terms 'toxic words', which actually mean something completely different from what is apparently implied. Thus, 'planning' is a mechanism for normalising people and ensuring conformity; identifying 'resources' is an excuse for desecrating nature; 'poverty' is an invention for undermining the values of traditional societies; and the application of 'science' is too often a justification for violence against indigenous peoples and their land. There is undoubtedly a degree of dramatic licence in this caricature, but it does nevertheless reveal the oppressive nature of the political and economic relationship between states at the opposite ends of the spectrum of prosperity.

World-systems analysis

World-systems analysis is a model, devised by Immanuel Wallerstein and elaborated in a series of major books published in the 1970s and 1980s, which attempts to draw together all the diverse threads in the debate about the nature of development into a single explanatory model (Wallerstein, 1974, 1979, 1980, 1983, and 1984). The model has assumed a particular importance in political geography, because it provided the analytical framework for much of the seminal work by Peter Taylor, including *Political Geography: world-economy, nation-state and locality*, probably the most influential textbook on political geography to appear in recent years (Taylor and Flint, 1999).

The core of Wallerstein's argument is that there have only ever been three basic ways in which societies have been organised to sustain and perpetuate the key processes of production and reproduction. What he terms the *reciprocal-lineage mode* describes societies that are mainly differentiated on the basis of age and gender and in which exchange is purely reciprocal. It is a model of economically simple, pre-feudal, and pre-industrial societies, that were for the most part highly restricted in their geographical range. They struggle to survive at all in the modern world, only maintaining a tenuous hold in some of the desert regions of southern Africa, and the tropical rainforests of South America, Asia, and Africa.

The *redistributive-tributary mode* describes societies that are class-based, with production carried on by a large majority of agriculturalists and paying tribute to a small ruling class. It is the classic conception of pre-industrial feudalism and was dominant in large parts of the world in what in Europe is known as the early modern era. The princes and maharajas of India, the emperors in China and Japan, as well as the petty rulers throughout Europe, were all part of this widespread system.

The *capitalist mode* is also class-based, but crucially is distinguished by ceaseless capital accumulation. The logic of the market dominates economic thinking and prices and wages are determined through the mechanisms of supply and demand. It is the mode of production that has come to define the modern world economy and it has systematically swept away, or at the very least marginalised, the two earlier modes.

Wallerstein contends that these three basic modes of production have, in their turn, resulted in three distinct types of society: mini-systems, world empires, and world economies. There have been innumerable mini-systems that have come and gone in the course of human history, and vast numbers of the misleadingly named world empires, going back throughout recorded history. To be more precise, the world empires actually refer to semi-closed economic and political systems, dominated by class-based hierarchies, and inhabiting a more or less discrete world of their own. In contrast, there has only ever been one world system, the capitalist world economy, which first emerged in Europe about the middle of the fifteenth century and, over the ensuing 350 years, spread to dominate the whole world. It is still all-powerful today, despite undergoing radical internal restructuring.

The key message of this analysis is that there can be no meaningful study of social, economic, and political change that proceeds on a country-by-country basis. It must incorporate the single society that is the world system. In other words, globalisation has been a fact for nearly 500 years and the inequalities built into it are systemic, not transitory, though their precise distribution is in a constant state of flux.

The structure of the world system is dominated by a single world market, but it also has a multi-state political framework. Within this system, no one state is ever able to dominate completely and certainly not in perpetuity. The more bombastic the claims to be eternal, such as Adolf Hitler's boast that the German Third Reich would last for a thousand years, the more short-lived they have tended to be. There is a constant political competition between states and it is this which gives economic decision-makers the leeway to manoeuvre and to look for new opportunities to increase their capital accumulation.

The world system can roughly be divided into three. At one end of the spectrum are the developed, industrialised states, forming the core. At the other are the largely non-industrialised, less developed states that have little to offer, other than their labour and supplies of raw materials that can find a market in the industrialised world. It is they that constitute the periphery. Between the two extremes is the semi-periphery, a highly politicised transition zone, where most of the movement occurs.

Political categorisation of the world in this way is replete with loose terminology, but the world system model has helped clarify some of the fundamental processes at work. For a generation, in the second half of the twentieth century, the term the Third World was used to designate those non-aligned states that resisted taking sides in the Cold War division of the world. It was also used as a piece of shorthand to describe the embattled territory between the two superpowers (Sachs, 1992). With the demise of the Soviet Union the term Third World has lost much of its meaning and the territory it represented is rapidly being reconstituted in a different way. A number of states, particularly in Asia, have seized on the opportunity to assert themselves economically and politically from within the world system's semi-periphery and to re-emerge as a new third world. Led by China, India and Indonesia, they are increasingly challenging the automatic dominance of the traditional developed countries in the core.

At the same time, what Manuel Castells (1997) has described as the Fourth World, has also begun to take shape. These are countries and regions that appear to be almost beyond the pale and appear to be in danger of falling outside the dynamic scope of the world system altogether. Many states in central Africa are now so poor and so ravaged by malaria, AIDS, and a host of other mortal diseases, that they offer little hope of trade and capital accumulation to the rest of the world, and are also too weak to generate any economic dynamism from their own resources (Kearns, 1996). The challenge for the globalised world is to find ways of preventing the hegemony of poverty and disease from creating a new, permanent, exclusion zone within the world system.

Key themes and further reading

The time–space compression has led to increasing integration and cross-fertilisation across the world, a process referred to as globalisation. All aspects of peoples lives have become more and more internationalised, a process that has been stimulated and encouraged by liberalised trade within the framework of international institutions like the OEEC, and its successor the OECD. Many firms now spread their operations across several national borders, and the largest (TNCs) appear to operate almost above the state system and outside its control. This enables them to bypass environmental and other restrictive legislation, though sometimes with disastrous consequences for the local population. In other cases, the internationalisation and globalisation of economic activity is driven by the search for raw materials, such as oil, sufficient supplies of which to satisfy the needs of industrialised countries cannot be found from domestic sources. Developmentalism argues that less developed countries, which have little to offer other than their labour and raw materials, will steadily improve their relative economic position over time, but this counsel of hope is now widely rejected. World-systems analysis postulates that the global economy has progressed through three stages to the modern world with just one world system. Within that system there is three-fold division in terms of economic power, with relatively little movement between the different levels. Indeed, there are some who postulate a Fourth World of states that are trapped in perpetual poverty at the bottom of the pile.

There are a number of excellent surveys of the progress of globalisation written by geographers. Two of the best are: *Global Shift: the transformation of the world economy* by Peter Dicken (1998), which has deservedly run to several editions; and *Mastering Space: hegemony, territory and international political economy* by John Agnew and Stuart Corbridge (1995), which has been very influential in making geographers reassess their view of the economic and political structure of the world economy at the turn of the twenty-first century. Another excellent survey of the world economy and the way it is evolving is *The Geography of the World Economy* by Paul Knox and John Agnew (1998). The most accessible introduction to world systems analysis is still *Political Geography: world-economy, nation-state and locality* by Peter Taylor and Colin Flint (1999).

12 International government and the modern state

Our country is the world – our countrymen are all mankind.

(William Lloyd Garrison, Prospectus of *The Liberator*, 15 December 1837)

The United Nations and the birth of world government

The interdependence and common purpose of humankind has been a recurrent political theme since at least the end of the eighteenth century, but two world wars and the growing globalisation of the world economy led to increased political pressure in the course of the twentieth century for effective governmental institutions that transcended national boundaries and interests. As has already been explained in Chapter 9, the first steps in this process met with very limited success, both in terms of promoting political stability, and in terms of opening up the world economy and reducing national protectionism. The League of Nations was unable to prevent the outbreak of the Second World War, the most destructive conflict ever seen in terms of loss of life, and a myriad of bilateral national agreements were strangling world trade.

The UN has been the bold and ambitious response to the challenge of a structure for world government, though still far from perfect and with major failures to set alongside its undoubted successes. The UN Charter was originally signed by 51 countries in 1945 and since then membership has grown steadily and, in 2004, stands at 185, representing all the formally recognised states in the world.

Since its inception, the UN has developed into a huge and somewhat unwieldy body, encompassing a wide range of very different international organisations under its umbrella (Figure 12.1). There are six

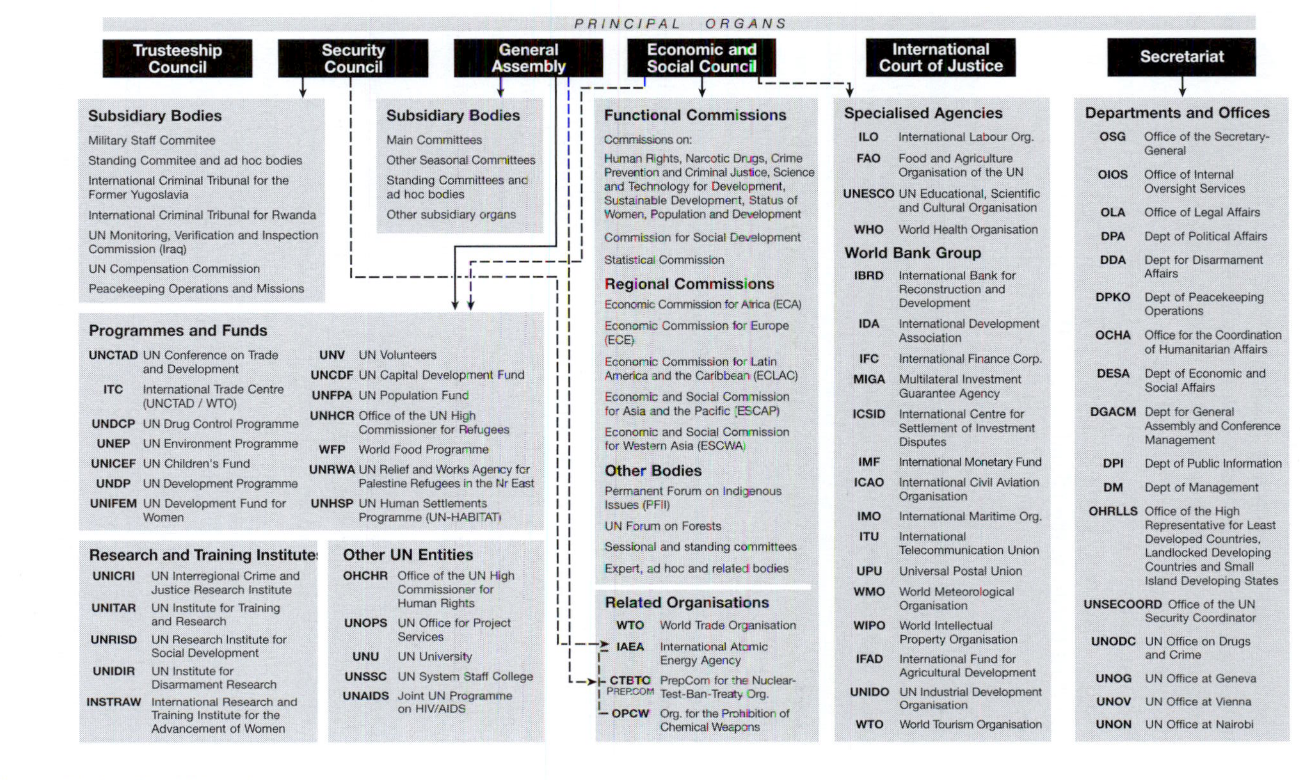

Figure 12.1 ***The UN system***

Note: solid lines from a Principal Organ indicate a direct reporting relationship; dashed lines a non-subsidiary relationship.

major divisions: the Trusteeship Council, the Security Council, the General Assembly, the Economic and Social Council, the International Court of Justice, and the Secretariat.

The *Trusteeship Council* was incorporated into the original UN Charter to oversee the progress of territories administered in trust by its member states towards full independence (Johnston *et al.*, 1988). It consists of the five permanent members of the Security Council – China, France, the Russian Federation (formerly the Soviet Union), the UK, and the USA – and it effectively completed its work in November 1994, when the last trust territory, the island state of Palau in the western Pacific Ocean, became independent and joined the UN in its own right.

The *Security Council* has eleven members, six appointed by the General Assembly and the other five the permanent members that form the Trusteeship Council. The main task of the Security Council is to maintain peace and security at an international level. It has a number of sanctions at its disposal, ranging from financial penalties, to trade embargos, and, as a last resort, military force. Any action must have the assent of seven of the eleven members, including that of the five permanent members. If military intervention is determined upon, then the Council can ask any of the members of the UN to contribute armed forces and other logistical support. The Security Council has a number subsidiary bodies, including all peacekeeping missions, the International Criminal Tribunal for the former Yugoslavia, the International Criminal Tribunal for Rwanda, and the Monitoring, Verification and Inspection Commission to search for evidence of weapons of mass destruction in Iraq. The International Atomic Energy Authority, a subsidiary of the Economic and Social Council, also has to report both to the Security Council and the General Assembly.

The *General Assembly* includes all the member states of the UN and is charged with discussing and deciding on issues of international peace and security. It can also make any recommendations it sees fit to promote international peace, as well as economic and social cooperation, and the promotion of human rights. It meets regularly and a two-thirds majority is required for any vote to be passed. The General Assembly has ten programmes and funds under its direct control, all of which have a seminal impact on the formulation and implementation of global policy. UNCTAD, the UN Conference on Trade and Development, has taken a leading role in deregulating world trade and promoting greater economic opportunities for all member states. The UN Drug Control Programme is at the centre of the worldwide struggle to control drug-related crime.

UNEP, the UN Environment Programme, is deeply involved in raising awareness of the threats associated with global environmental pollution and climate change. UNICEF, the UN Children's Fund, leads the way in exposing and combating the suffering and exploitation of children across the globe. The UN Development Programme oversees and coordinates development initiatives. The UN Population Fund monitors and advises on the demographic issues arising from the inexorable growth of the world's population. UNHCR, the Office of the UN High Commissioner for Refugees, actually started work before the UN Charter was signed and ever since has attempted to inject some order and humanity into the intractable problems facing displaced people. The World Food Programme provides essential support for the UNHCR, and other programmes, by ensuring that essential foodstuffs reach those who are starving and most in need. The UN Relief and Works Agency for Palestine Refugees in the Near East, as its name suggests, is a special agency that works to relieve the unique problems facing the Palestinian people in that region. UN-Habitat, the UN Human Settlements Programme, struggles with the huge challenge of providing adequate housing in urban and rural areas for a world population that is constantly on the move, both between countries and within them.

These are just the major programmes. There are a number of others, as well as research and training institutes and, all together, they represent a huge investment on the part of the world community to reduce human hardship of all kinds. Naturally, some programmes are more successful than others and none are successful all the time, but, taken as a whole, they represent a massive support mechanism for helping individual states to cope with problems that threaten to overwhelm their individual national resources.

The main task of the *Economic and Social Council* is to seek to improve the economic and social well-being of those living in the member states. Its brief covers health, education, economic, social and cultural issues, and the promotion of the position of women in the world. It supports a wide range of functional and regional commissions, as well as more than twenty specialised agencies, covering all aspects of the world economy (the World Trade Organisation, the World Bank, the International Monetary Fund), as well as education, science, and cultural development (UNESCO), and many other areas.

The *International Court of Justice* is the main judicial body of the UN and all members have to agree to abide by its decisions. The court consists of fifteen members and no country may have more than two members at any one time. Its existence is an important step forward in ensuring that international decisions are actually implemented. But it still suffers from serious weaknesses. The court is almost entirely dependent on national governments to enforce its rulings, and many states have been reluctant to refer matters to the court, because they fear it may not find in their favour.

The *Secretariat* is headed by the Secretary-General, who is appointed by the General Assembly, and is the body that runs the whole complex and widespread organisation that is the UN in the twenty-first century. While the UN does not have the power to force individual member states to follow its recommendations, it does exert a considerable moral authority, which makes rejecting them a serious political step. Furthermore, in the space of little more than half a century, the world has moved from a position where it had failed disastrously with the League of Nations to create a viable world political order, to one where the absence of the UN and its many agencies would be inconceivable, despite its undoubted shortcomings.

Antarctica

No issue illustrates better the strengths and limitations of the UN than its role in brokering a globally acceptable and sustainable political future for Antarctica, the last great unsettled wilderness on earth with over 70 per cent of the world's freshwater locked up in its ice-sheets (Schram Stokke and Vidas, 1996). The continent was first discovered by the British mariner and explorer, James Cook, in 1774, but there was no permanent presence there until 1943, when the UK established a base to provide reconnaissance and meteorological information for the south Atlantic. The UK also made the first unilateral claim to sovereignty over the Antarctic landmass, the British Antarctic Territory, when it defined a wedge-shaped area in the north-west of the continent, stretching from the coast to the South Pole, as well as further out into the south Atlantic to include the South Sandwich Islands and South Georgia. By the early 1950s, six other states – Norway, Australia, France, New Zealand, Chile, and Argentina – had also lodged similar wedge-shaped claims, with those of Argentina, Chile, and the UK overlapping each other, so that over 80 per cent of the continent was technically spoken for, though none of the claims was recognised by other, non-claimant states in the world (Figure 12.2).

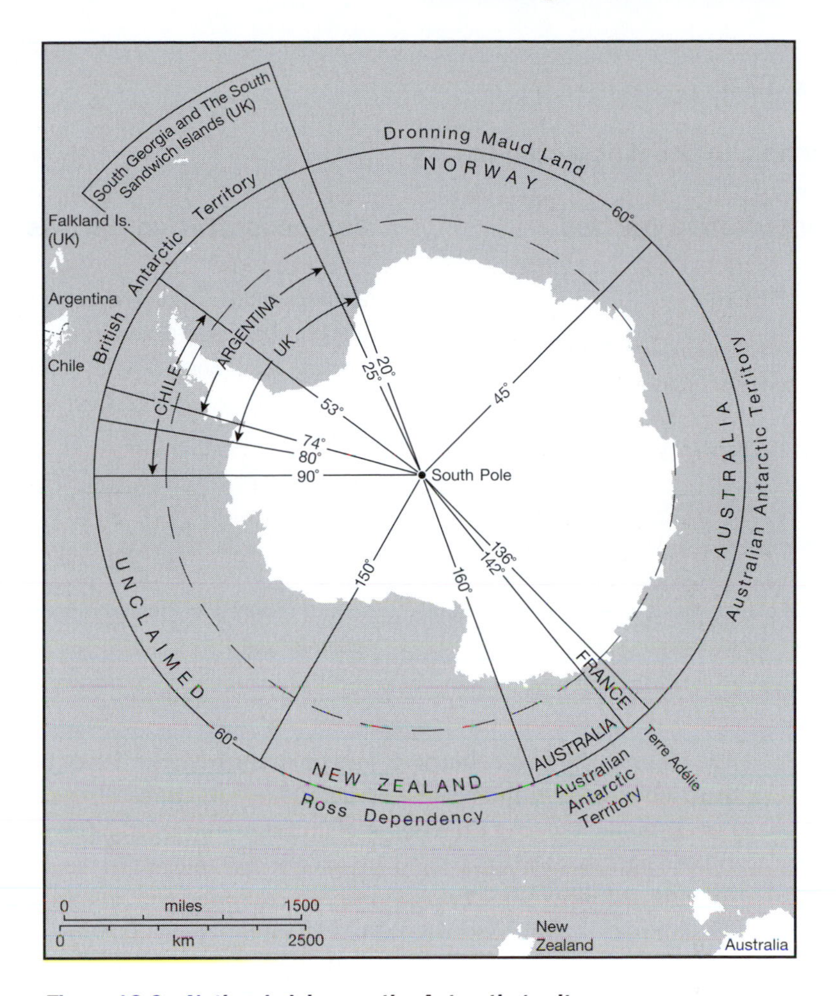

Figure 12.2 ***National claims on the Antarctic territory***

Politically, it was clearly an unsustainable situation and, following on the success of the twelve-nation international scientific collaboration in to all aspects of the Antarctic environment during the International Geophysical Year in 1957–8, it was decided to draw up an Antarctic Treaty, monitored and administered by the UN. The treaty was published in 1959 and came into force in 1961 (Box 12.1). It covered the whole of the area south of 60° latitude south and its objectives were: to keep the continent permanently demilitarised and to establish it as a nuclear-free zone, thus ensuring that it is used only for peaceful purposes; to promote international scientific cooperation; and to put on one side all existing disputes over territorial sovereignty (Dodds, 1997).

Box 12.1

Parties to the Antarctic Treaty

Consultative parties

Claimants

Argentina*, Australia*, Chile*, France*, New Zealand*, Norway*, UK*.

Non-claimants

Belgium*, Bulgaria, Brazil, People's Republic of China, Ecuador, Finland, Germany, India, Italy, Japan*, Netherlands, Peru, Poland, Russia*, South Africa*, South Korea, Spain, Sweden, USA*, Uruguay.

Non-consultative parties

Austria, Canada, Colombia, Cuba, Czech Republic, Denmark, Greece, Guatemala, Hungary, North Korea, Papua New Guinea, Romania, Slovakia, Switzerland, Turkey, Ukraine.

Note: * denotes an original signatory of the Antarctic Treaty.

The UK was the first country to sign up to the treaty in 1961, but there are now forty-three signatories, divided into three distinct groups. First there are the seven territorial claimant countries, four of which are in the southern hemisphere and have a riparian interest in the south Atlantic, the other three being European countries with a long-standing commitment to polar research and exploration. Second, there are twenty non-claimant countries, mostly the major industrialised nations, identified as consultative parties, who must be involved in any changes to the status of the treaty. Third, there are sixteen non-consultative parties, states that wish to register an interest in the future of Antarctica, but recognise that they will never be in a position to exert a major influence on what is actually decided.

Within the framework of the treaty, five further measures have been adopted to protect the Antarctic environment and its flora and fauna, including that of the surrounding seas, and to regulate any future mineral exploration and development. In practice, the last of these measures, the Protocol on Environmental Protection to the Antarctic Treaty, which was agreed in 1991 and enacted in 1998, essentially superseded the others and combined them into a single document. In many ways, these subsequent additions are more important than the original treaty itself, because they provide for regimes to control fishing and the taking of marine mammals in the seas around Antarctica, which at present are the only exploitable natural resources.

The treaty can remain in force indefinitely, though there was provision for a review after 30 years, but in 1991 none of the consultative parties wished to see it renegotiated. Clearly they believe it is working well and in their interests, but part of their lack of enthusiasm for any change is the growing clamour from countries that are not party to the treaty. Ever since 1982 the future of Antarctica has been raised regularly in the UN General Assembly with calls for the whole continent to be designated a 'Global Common' or 'Heritage of Humankind'. These calls have come from non-treaty countries in the developing world, which feel, with some justification, that the continent ought to be a global resource and that, if it is not, they have been frozen out of any long-term economic benefits arising from development in Antarctica. Not surprisingly, the countries that are already parties to the treaty have opposed any suggestion that their privileged position be threatened in this way and, thus far, have successfully resisted any change (Suter, 1991).

The role of the UN in the recent history of Antarctica has been somewhat ambivalent. On the one hand, it has strongly supported the objectives of the treaty, in so far as they ban all military activity and nuclear testing, put strict limits on any exploitation, and support scientific research. On the other hand, it must also try to fight the corner of the very large number of small states that so far have been excluded from any say in the future of the continent.

Managing the global market

Despite the scale and huge scope of its activities, the UN is far from being a complete answer to the need for global institutions, especially when it comes to international finance. For more than two decades after the organisation was founded, the USA, or more precisely the US dollar, was the foundation on which trade in the non-Communist world relied, but, in the 1970s, this whole edifice began to disintegrate, partly as a result of economic weakness in the USA, and partly as a result of the growth in the sheer volume of trade in the world (Barry, 2001).

In the face of an economic crisis, fuelled further by the sudden sharp increase in the price of oil on the world market in 1973, the six major capitalist countries in the trade-dependent maritime world (USA, Federal Republic of Germany, France, UK, Japan, and Italy) in 1975 formed themselves into an ad hoc group, known as the G6, to create a forum for regular meetings of their finance ministers to try to coordinate their macro-economic policies. At the insistence of the USA, Canada was invited to join in 1976 and, what was now the G7, began to exert a powerful influence on all aspects of economic policy. Indeed, broadly in the name of economic policy, the group's pronouncements ranged widely into other areas of policy-making as well. For instance, it roundly condemned the Soviet invasion of Afghanistan in 1979, as well as playing a leading role in attempts to deregulate world trade within the general framework of the UN-sponsored General Agreement on Tariffs and Trade (GATT).

In the ensuing years the G7 has become an increasingly powerful voice and one that has been further strengthened since the collapse of the Soviet Union in 1989. In 1991, the group invited the newly formed Russian Federation to join it as an observer, subsequently inviting it to become a full member in 1998 and thus changing the group into the G8. In 1999, in the face of growing criticism at its apparent exclusiveness and dominance, the G8 invited twelve other countries with emergent industrial economies to join with it in a subsidiary forum, the G20 (Box 12.2), to involve them more closely in discussions about the direction of global economic policy; this consultative forum has subsequently been enlarged to form the G33.

Box 12.2

Membership of the G8 and G20

The members of the informal groups, G8 and G20, are the finance ministers and central bank governors of the following states:

G8

Canada, France, Germany, Italy, Japan, Russia, UK, USA.

G20

Includes the G8 countries, together with Argentina, Australia, Brazil, China, India, Indonesia, Mexico, Saudi Arabia, South Africa, South Korea, Turkey, and the EU in its own right.

The cornerstone of world trade policy is the World Trade Organisation (WTO) and its precursors. After the Second World War, attempts to create an International Trade Organisation, led by the USA, failed and the GATT was created as a temporary measure in its place. Starting with 23 members in 1947, it grew rapidly and by the early 1990s had a membership in excess of 120. GATT operated through a series of rounds of multilateral negotiations, the aim of which was to reduce progressively tariffs and other barriers to free trade across the world. From its inception, it has been guided by two basic principles. The most-favoured nation principle insists that any trade concession agreed bilaterally between two countries must apply equally to all the other members of the GATT. The national treatment rule requires that all imported goods are traded in exactly the same way as domestically produced goods. A third principle was agreed in 1965, which allowed preferential access for goods produced in developing countries to markets in the developed, industrial world, though there were some important exceptions, associated especially with clothing and textiles, and there is some cynicism about how much improvement in access was actually achieved (Hoekman and Kostecki, 1995).

By 1994, eight rounds of negotiations had been completed within the framework of GATT and more than 90 per cent of world trade in raw materials and manufactured goods was covered by the agreements. The Uruguay Round, the last to be completed in 1994, was by far the most ambitious. It extended the scope of GATT to include, for the first time, agriculture, textiles, and clothing. It also concluded preliminary agreements on trade in services, intellectual property, and investment, thereby encompassing whole new areas of economic activity, which account for a growing proportion of the world economy.

The success of the Uruguay Round also provided an opportunity to bring an end to the provisional status of GATT and to replace it with a permanent body. The WTO came into existence in 1995 and since then has continued the work of trying to eliminate restrictions on trade of all kinds. GATT had been very successful at eliminating overt restrictive measures, like tariffs and import duties, but much less so in getting to grips with other forms of trade distortion, such as quotas on the amounts of goods that could be imported and legal requirements which prevented goods and services from having access to certain national markets. For instance, a refusal to recognise professional qualifications gained in another country can very effectively prevent doctors, lawyers, accountants, and many others from practicing outside their own national jurisdictions.

The WTO has begun to address the detail of these issues, as well as continuing to try to eliminate tariffs and other, more obvious, limits on free trade. In 2001, it initiated the Doha Round, an attempt to reduce still further the very severe obstacles that most of the major developed countries continue to place in the way of trade in agricultural products. Heavy agricultural subsidies in the developed world mean that they are able to produce foodstuffs at artificially low prices and, as a result, developing countries, whose main exports are such primary products, are unable to penetrate the lucrative markets in the developed world, especially in North America and Europe.

Progress in the Doha Round has been very slow and the failure of the WTO to act more quickly and effectively to address the widespread discrimination through the barriers to trade between the developed and the developing world has led to widespread criticism. Protests have been growing, and are becoming increasingly violent, in the face of the apparent lack of will in the organisation to truly create a more equitable environment for a trading system that supports the interests of all the countries in the world, and not just those that are already industrialised and affluent (Chossudovsky, 1997).

Regional responses

By no means all the responses to globalisation have been global in extent; regional initiatives have been equally important, in some cases transforming the geopolitical context of political decision-making. Regional alliances in support of furthering national economic interests and bolstering national defensive capabilities have, of course, long been a feature in the world's political landscape, but since the middle of the twentieth century some have begun to assume more concrete form, with their own institutions, and economic, political, and social infrastructure.

In the forefront of such developments is the EU, a regional political initiative to which frequent reference has already been made, that has transformed relations between European states and, in a period of little more than half a century, has remade the political architecture of the whole continent (Blacksell and Williams, 1994; Williams, 1994). It began life as a core group of six, largely industrial states in mainland western Europe (Belgium, France, Federal Republic of Germany, Italy, Luxembourg, and the Netherlands), which were trying to rebuild their economies in the wake of two disastrous world wars in the space of less than forty years, first as the European Coal and Steel Community (1952), and then as the European Economic Community (EEC) (1956). The goal was to create a customs union, with free trade amongst themselves and a common external policy for trade with the rest of the world, as well as to develop common policies in specific areas, such as agriculture and transport.

Throughout the 1960s much was achieved, though progress was very erratic. Nonetheless, the EEC was secure enough to feel able to accede to the requests of other West European countries to be allowed to join. Denmark, Ireland, and the UK joined in 1973, but the expansion coincided with the time that the US-led period of sustained economic growth, that Western Europe had enjoyed since the early 1950s, was beginning to break down. What now had been renamed the European Community (EC), experienced a very difficult few years in the mid-1970s, both economically and politically, with even its long-term future being thrown into question.

The European Community managed to weather the storm and, once it re-emerged, it was well on the way to becoming a substantially different political entity. In 1979, the first direct elections to the European Parliament took place and marked the beginning of a long process of giving the EC a political voice and presence, independent of its member states. In 1981, Greece became a member, followed by Portugal and Spain in 1986, all three joining less for economic reasons than to underpin their newly elected, and fragile, democratic governments. The EC was manifestly becoming a political force in its own right and this trend was accelerated even further by developments after 1990.

First of all, the EC itself took a much stronger lead than hitherto in moving from trade liberalisation to the creation of a deeply embedded single market across all its member states. The Maastricht Treaty in 1994 not only reformed the EC as the integrated European Union (EU), it also committed it to becoming a single market with its own currency, the euro, as well as its own social and environmental agenda (Wise and Gibb, 1993). At the same time, it had to respond to dramatic political upheaval, sparked by the collapse of the Soviet Union. In 1990 it effectively welcomed its first member from the former Communist bloc, when Germany was reunified and what had been East Germany became included in the union. It was immediately clear that this was likely to be the beginning of a new phase of enlargement in support of a new political agenda, and that the EU of twelve members was unlikely to be able to rise to the economic challenges that this would bring with it. Other wealthy states in Western Europe were encouraged to join and, in 1995, Austria, Sweden, and Finland became members, thus easing the burden of integrating a group of much poorer former Communist states from Eastern Europe. There then followed a period of intense negotiation, which culminated in the EU accepting ten new members in 2004, eight former Communist states in Eastern Europe (Estonia, Latvia, Lithuania, Poland, the Czech Republic, Slovakia, Hungary, and Slovenia), together with Cyprus and Malta, two small Mediterranean island countries.

The twenty-five members of the EU now form a very significant global political entity that has a population of 456 million, more than half as big again as that of the USA, with negotiations for further new members (Bulgaria, Romania, and Turkey) already in train. Although it is not a new state in the conventional sense, it does have many of the trappings of statehood: a democratically elected parliament, its own currency, which has been adopted by the majority of its member states, and a constitution that is drafted and under discussion.

Whether the EU will be a model for other regions of the world remains to be seen, but there have certainly been putative attempts to follow its example. NAFTA, the North American Free Trade Area, ASEAN, the Association of South-East Asian Nations, and LAFTA, the Latin American Free Trade Association, all represent regional free trade agreements, but none has so far begun to assume the separate institutional substance of the EU.

Regional defence alliances have been commonplace throughout history, but they have usually been ephemeral and have not evolved to create any institutions, independent of the individual member states involved. NATO, the North Atlantic Treaty Organisation, is an exception (Blacksell, 1981). Founded in 1949, the alliance comprised the USA and Canada and a steadily expanding group of Western European countries, committed to containing the threat of Communist expansionism during the Cold War. The key to the alliance was a commitment that an attack against any one of the signatories would be treated as an attack against them all, and would spark an appropriate military response. By 1982 there were fourteen West European countries in the alliance and NATO had developed a very substantial separate institutional presence, based around its headquarters in Brussels. It had also evolved a system of joint commands, so that NATO forces were integrated across national lines in a way that remains highly unusual.

The collapse of the Soviet Union should, theoretically, have marked the end of the alliance, as the military threat it had been set up to contain had effectively disappeared, but predictions of its imminent demise proved premature. NATO had become so integral to the whole defence structure of Western Europe and the North Atlantic region that, rather than fading away, it has been transformed into a vehicle for integrating the former Communist states of eastern Europe into a unified defensive alliance. After the unification of Germany in 1990, it oversaw the withdrawal of Russian troops from what had been East Germany. Poland, Hungary, and the Czech Republic were admitted as members in 1998 and, in 2002, they were followed by Bulgaria, Estonia, Latvia, Lithuania, Romania, Slovakia, and Slovenia. It is also highly likely that more former east European states will join in the future.

The inclusive, multinational approach adopted by NATO towards the former Communist states of eastern Europe has almost certainly been crucial in smoothing the way for their peaceful transition to democracy. It also stands in sharp contrast to the fate of some of the former Soviet republics in the Caucasus, where vicious civil wars have all but stopped democracy taking root.

Human rights

One aspect of the tortuous institutional response to a more internationalised world which must not be forgotten is the struggle for universal standards in human rights (Smith, 1994). The First French Republic, which emerged from the political anarchy of the French Revolution in 1791, and the US Bill of Rights, which was enunciated in the same year, both set standards for human rights and freedoms and in the twenty-first century all states incorporate into their constitutions or systems of governance laws regulating the relations between those in power and citizens. However, although some, such as the right to life, are almost universal, others, such as freedom of speech and the right to vote in a democratic society, are not. Indeed, it is sometimes argued that to attempt to impose a set of universal rights is actually not desirable, as it will stifle legitimate differences between societies.

Nevertheless, the growth in the number and importance of international institutions has led to an ever greater insistence on international standards for human rights. The first Geneva Convention which attempted to put limits on the conduct of war and the treatment of prisoners was signed in 1864 and has been extended through a series of further conventions and protocols ever since. The UN adopted the *Universal Declaration of Human Rights* in 1948 and in 1966 adopted the *Covenant on Civil and Political Rights* and the *Covenant on Economic, Social and Cultural Rights*, both of which are binding on its member states and provide an effective global framework for human rights. Earlier, the Council of Europe, created in 1949, was an attempt to generate common basic standards of behaviour across Europe and, although it only encompassed non-Communist countries during the Cold War, it has expanded rapidly since 1990 and now has forty-six members from across the whole of the continent, including the Russian Federation. The greatest achievement of the council has been the *European Convention on Human Rights*, which most of its members have accepted and incorporated into their constitutions. It sets a general standard for human rights, which can be tested in the council's legal forum, the European Court of Justice, effectively providing a supreme court for deciding human rights issues. The UN has also taken action to set up formal institutions to adjudicate on violations of human rights mainly through the International Court of Justice, but also by establishing International War Crimes Tribunals to deal with human rights and genocide in Rwanda and the former Yugoslavia.

Key themes and further reading

The growing recognition of the essential interdependence of all humankind and the frequency and complexity of interaction at a world scale increasingly require institutions to underpin them. The opportunities and limitations are well illustrated by the way in which the UN has been organised to meet these challenges. The Antarctic Treaty is the most ambitious attempt to place an area of the globe of continental scale above national politics and competitiveness. The realisation of the ideal of a global marketplace has also led to institutionalisation through GATT and, subsequently, the WTO. Regional organisations for military cooperation have been commonplace for centuries, but they too have begun to assume stronger identities in the modern world through bodies such as NATO. Regional economic cooperation is also increasingly common, notably in the form of the burgeoning EU. The greater contact between peoples across the globe has brought with it demands for common minimum standards of behaviour and the recognition of human rights. This applies as much to the way governments treat their own peoples, as it does to their behaviour towards each other.

The recent institutional development of the world trading system is considered at length by B. Hoekman and M. Kostecki (1995) in *The Political Economy of the World Trading System: from GATT to WTO*. For a study of the evolution of the EU from a purely economic arrangement to a much more sophisticated and complex transnational socio-economic organisation, *Single Market to Social Europe* by Mark Wise and Richard Gibb (1993) provides an excellent introduction. The political geography of Antarctica as a continent under international, rather than national, control is dealt with in depth by Klaus Dodds (1997) in *Geopolitics in Antarctica: views from the Southern Ocean rim*. Turning to human rights, this whole field, and much more, is surveyed from a geographical perspective by David Smith (1994) in *Geography and Social Justice*.

13 Conclusion

The shape of things to come

Now each man was a nation in himself, without mother, father, brother.
(Derek Walcott, *Omeros*, 1990, Faber & Faber, London, chapter 28(1))

Adjusting the focus

In many ways political geography is a perfect metaphor for the postmodern era. People have constantly tried to reshape politically the world in which they live in their own image and attempts to try to limit the extent of the process of transformation have always ultimately failed. The process has also been unpredictable, with new patterns and groupings emerging and flourishing, apparently defying conventional logic and expectations. A generation ago, at the height of the Cold War, few would have predicted the current extent of the EU, encompassing most of the former Communist-controlled states in central and eastern Europe and with active plans to include yet more in the near future. Equally, it was certainly never envisaged that over much the same period the continent of Africa would change politically from a series of large European colonial territories supplying cheap raw materials to the developed industrialised world, into a network of independent states, struggling against poverty and disease to survive in the globalised world economy of the twenty-first century.

In many instances, such momentous changes actually are unforeseen and unplanned, but others owe much to political artifice and design and it has been the contribution of political geography to throw at least a little light on the dynamics behind these changes. In the process, it has also undoubtedly promulgated a very skewed view of the world, focusing on the achievements of some groups which conform to accepted measures of success and significance, while virtually ignoring others.

Historically, political change is viewed as having been driven by men, as opposed to women, and, what is more, men in recognised positions of power, internationally, nationally, regionally, and locally. It is, however, a very partial view and one that is increasingly contested and challenged, especially by women in geography (Domesh, 1991). More than a decade ago, Kofman and Peake (1990) set out what they called a gendered agenda for political geography, with the explicit intention of shifting the focus of debate away from male-dominated orthodoxy, to a more subtly layered view of political change. Their challenge was partially accepted and a new genre of research and writing did begin to emerge, highlighting how differently development and political change had impacted on different groups in society in different parts of the world (Momsen and Kinnaird, 1993), but the inroads were small and there is still much more to do to redress the balance of the debate in political geography in favour of the less powerful and less strident elements in society.

More recently, political geography has still not engaged fully with the challenges posed by the internet and other dramatic advances in information technology (Castells, 1997). Communication is now global, virtually instantaneous, almost entirely unregulated by government, and cheap. This IT revolution is having a fundamental impact on the business of politics and on political geography. There has been a wholesale assault on political boundaries of all kinds, not simply because they can now so easily be transcended, but also because those wishing to do so can alter their line of attack almost at will and are, therefore, almost impossible to pin down, control, and eliminate.

The best known of the international terrorist organisations posing a direct threat to the authority of established governments is Al Qaeda, an organisation founded in 1988 in the wake of the successful guerrilla war waged by the Muslim *mujahideen* to drive the Soviet army out of Afghanistan. Subsequently, it has developed into a worldwide terrorist organisation, aimed at removing pro-American influence and control from the Muslim world, its most spectacular coup being the devastating attack on the World Trade Center in New York on 11 September 2001.

Initially, Al Qaeda was largely based in Afghanistan, but since the American-led invasion in 2002 and the overthrow of the Taliban government in that country, it has transformed itself into a highly dispersed global network of terrorist cells with bases in at least twenty-six countries. The network is coordinated using highly sophisticated IT systems and technology and operates almost entirely outside the established political networks.

The impact of this, and other less successful and extensive political networks, has been seriously to destabilise the world order, leading the USA to declare a global war on terror, with an enemy that is almost impossible to define in conventional spatial terms. Clearly delineated geopolitical theatres of influence, in the sense that they were defined by Mackinder and others throughout the twentieth century, are increasingly irrelevant, making it imperative that political geographers devise new ways of representing the sway of political power.

Ultimately, however, the most important challenge for political geography is to ensure that it absorbs and takes account of the major regional shifts in political power that are currently occurring in the world. Inevitably the current literature is heavily biased towards Anglo-America and Europe, with the global influence of the USA in particular undoubtedly dominating political discourse and the world political map in recent times. Whilst there is absolutely no immediate sign of that altering fundamentally in the short term, political geographers will need to watch carefully the rapid emergence of other potential super-states, with populations, geographic areas and, increasingly, the economic power to pose a real challenge to American dominance.

Pre-eminent amongst these will be the People's Republic of China, which with 1.3 billion is by far the most populous country in the world, as well as being the fourth largest in terms of area (9.6 million sq km). Even more significantly, its GDP stood at US$6.4 trillion in 2003, second only to that of the USA and growing at a rate that makes it likely that it will move into first place within a decade. Interestingly, the economic transformation of China has been achieved without the benefit of a market-led liberal democracy, which is usually cited as the keystone to the success of the USA. Elsewhere in the world, India, though somewhat smaller than China with a population of just over 1 billion, is growing even faster, having witnessed an increase of over 20 per cent in the decade up to 2001, while other states in south-east Asia, such as Indonesia, are also now major economic and political influences on the world stage.

It all adds up to a fundamental refocusing of global power. Just as North America grew to challenge and overtake Europe in the early years of the twentieth century, so China, India, and the rest of Asia look poised to do the same at the beginning of the twenty-first century.

Reappraising the state

One of the most surprising recent developments in political geography is the renewed enthusiasm for states as the fundamental building blocks of the political order. Commentators, like Francis Fukuyama who, in the wake of the collapse of the Communist system in the Soviet Union and its satellite states, could somewhat gleefully proclaim the ultimate victory of the free market economy in underpinning successful societies across the world, have become much more cautious in their assessments (Fukuyama, 1992 and 2004). Fukuyama now argues that the role of states and state institutions have been undervalued, as compared to measures of economic performance, and that successful markets are crucially dependent on states with what he terms 'good governance'. What exactly constitutes 'good governance' is not entirely clear, as success has stemmed from widely differing political models, from the modified totalitarianism of the People's Republic of China, to the monumental experiment in popular democracy that is modern-day India. Nevertheless, robust state structures are seen as crucial for stability and growth, and weak states as a political liability that can pose a threat way beyond their borders.

Strong and effective states, offering good governance, are also seen more and more as the only way to curbing the excesses of global corporate power (Wainwright, 2004). Governance at the world scale, in the shape of the UN and other international institutions, has made huge progress since the middle of the twentieth century, but remains at the most a partial solution, heavily dependent on the active support of individual states. It is they that have the power, and the crucial popular legitimacy, to deliver the infrastructure and services that societies require; it is also increasingly accepted that they can often do it better than private providers, or at the very least that they provide a very necessary control over the quality of what the private sector does provide.

Understanding these pivotal roles for the state is a key challenge for modern political geography, as it strives to explain the relationship between people and the political infrastructure that binds societies together. It reaffirms the centrality of political geography within geography, even if its role is now very different from that envisaged by Friedrich Ratzel when he first identified political geography as a core element in the discipline as a whole.

References

Agnew, J. (1996) 'Mapping politics: how context counts in electoral geography', *Political Geography* 15: 129–46.

—— (1997) 'The dramaturgy of horizons: geographical scale in the "Reconstruction of Italy" by the new Italian political parties, 1992–95', *Political Geography* 16: 99–121.

—— (2000) 'Territoriality', in Johnston, R. J., Gregory, D., Pratt, G., and Watts, M. (eds) *The Dictionary of Human Geography.* Fourth edition. Blackwell, Oxford, pp. 823–4.

—— (2002) *Making Political Geography.* Hodder & Stoughton, London.

Agnew, J. and Corbridge, S. (1995) *Mastering Space: hegemony, territory and international political economy.* Routledge, London.

Alvarez, A. (1924) *The Monroe Doctrine: its importance in the international life of the states of the New World.* Oxford University Press, New York.

Archer, J. C. and Taylor, P. J. (1981) *Section and Party: a political geography of American presidential elections from Andrew Jackson to Ronald Reagan.* John Wiley, London.

Bardi, L. (1996) 'Transnational trends in European parties and the 1994 elections of the European Parliament', *Party Politics* 2: 99–114.

Barry, T. (2001) 'G8 and global governance', *Foreign Policy in Focus* 6(27).

Bassin, M. (1987a) 'Imperialism and the nation state in Friedrich Ratzel's political geography', *Progress in Human Geography* 11: 473–95.

—— (1987b) 'Friedrich Ratzel 1884–1904', *Geographers: bibliographical studies* 11: 123–32.

Beckinsale, R. P. (1972) *Companion to Berkshire.* Spurbooks, Bourne End, Buckinghamshire.

Bell, M., Butlin, R. A., and Heffernan, M. (1994) *Geography and Imperialism, 1820–1940.* Manchester University Press, Manchester.

Blacksell, M. (1981) *Post-war Europe: a political geography*. Revised edition. Hutchinson, London.

—— (1982) 'Reunification and the political geography of the Federal Republic of Germany', *Geography* 67: 310–19.

—— (1998) 'Political parties', in Unwin, T. (ed.) *A European Geography*. Addison Wesley Longman, London, chapter 8, pp. 115–28.

Blacksell, M. and Brown, M. (1983) 'Ten years of Ostpolitik', *Geography* 68: 260–2.

Blacksell, M. and Williams, A. M. (eds) (1994) *The European Challenge: geography and development in the European Community*. Oxford University Press, Oxford.

Blacksell, M., Clark, A., Economides, K., and Watkins, C. (1990) 'Citizens' Advice Bureaux: problems of an emerging service in rural areas', *Social Policy and Administration* 24: 212–25.

Blacksell, M., Economides, K., and Watkins, C. (1991) *Justice Outside the City: access to legal services in rural Britain*. Longman, London.

Blacksell, S. and Phillips, D. R. (1994) *Paid to Volunteer: the extent of paying volunteers in the 1990s*. Third Series, Paper 2, The Volunteer Centre UK, London.

Block, R. (1980) 'Frederick Jackson Turner and American geography', *Annals of the Association of American Geographers* 70: 31–42.

Blouet, B.W. (1987) *Halford Mackinder: a biography*. Texas A. & M. University Press, Texas.

—— (2004) 'The imperial vision of Halford Mackinder', *Geographical Journal* 170: 322–9.

Boal, F. (1969) 'Territoriality on the Shankill–Falls divide, Belfast', *Irish Geography* 6: 30–50.

Bondi, L. (1997) 'In whose words?: on gender identities, knowledge and writing practices', *Transactions of the Institute of British Geographers* NS 22: 245–58.

Boudeville, J. R. (1966) *Problems of Regional Economic Planning*. Edinburgh University Press, Edinburgh.

Bowman, I. (1928) *The New World: problems in political geography*. Fourth edition, World Book Co., Yonkers-on-Hudson NY and Chicago IL (first published 1921).

Brezezinski, Z. (1997) *The Grand Chessboard: American primacy and its geostrategic imperatives*. Harper Collins, New York.

Brunn, S. D. (1974) *Geography and Politics in America*. Harper & Row, London.

Bullard, R. (ed.) (1993) *Confronting Environmental Racism: voices from the grassroots*. South End Press, Boston MA.

Burghardt, G. (1969) 'The core concept in political geography: a definition of terms', *Canadian Geographer* 13 (4): 349–53.

—— (1972) 'The bases of territorial claims', *Geographical Review* 62: 225–45.

Carr, E. H. (1945) *International Relations since the Peace Treaties*. Macmillan, London.

—— (1945, reprinted1968) 'The climax of nationalism', *Nationalism and After*. Macmillan, London, pp. 1–34.

Carson, R. (1962) *Silent Spring*. Houghton Mifflin, Boston MA.

Castells, M. (1977) *The Urban Question: a Marxist approach.* Edward Arnold, London.
—— (1997) *End of Millennium: the information age, economy, society and culture.* Blackwell, Oxford.
Charity Commission for England and Wales (2002) *Responsibilities of Charity Trustees.* Publication 3a, The Charity Commission, London.
Chazan, N. (1991) *Irredentism and International Politics.* Adamantine, Twickenham.
Cho, G. (1998) *Geographic Information systems and the Law.* Wiley, Chichester.
Chossudovsky, M. (1997) *The Globalisation of Poverty: impacts of IMF and World Bank reforms.* Zed Books, London.
Christopher, A. J. (1999) *The Atlas of States: global change 1900–2000.* Wiley, Chichester.
Citizens Advice (2003) *Annual Review.* NACAB, London.
Clark, G. L. and Dear, M. (1984) *State Apparatus: structures and language of legitimacy.* Allen & Unwin, Boston MA.
Cleverly, R. (2004) 'Definition of maritime space: charting the baseline – or not?' Unpublished paper delivered to the Special Session: GIS on Marine and Coastal Environments at the Annual Meeting of the Association of American Geographers, Philadelphia PA, 14–19 March.
Coe, N. M., Hess, M., Yeung, H. W., Dicken, P., and Henderson, J. (2004) '"Globalizing" regional development: a global production network perspective', *Transactions of the Institute of British Geographers* NS, 29: 468–84.
Cohen R. S. (1978) 'State origins: a reappraisal', in Claessen, H. J. M. and Skalnik, P. (eds) *The Early State.* Mouton, The Hague.
Cohen, S. B. (1964) *Geography and Politics in a Divided World.* Methuen, London.
—— (1982) 'A new map of global political equilibrium', *Political Geography Quarterly* 1: 223–41.
—— (1992) 'Policy prescription for the post-Cold War world', *Professional Geographer* 44: 13–15.
—— (2002) 'Earth and State, a study in political geography', *Political Geography* 26: 679–82.
Couper, A. (1978) *Geography and the Law of the Sea.* Macmillan, London.
Cox, K. (1973) *Conflict, Power and Politics in the City: a geographic view.* McGraw-Hill, New York.
Crampton, J. (1994) 'Cartography's defining moment: the Peters projection controversy 1974–1990', *Cartographica* 31: 16–32.
Crang, M. (1998) *Cultural Geography.* Routledge, London.
Crang, P. (1996) 'Displacement, consumption and identity', *Environment and Planning A* 28: 47–67.
Curzon of Kedleston, Lord (1907) *Frontiers: the Romanes Lectures.* Oxford University Press, Oxford.

Cutter, S. (1995) 'Race, class, and environmental justice', *Progress in Human Geography* 19: 107–18.

Dalby, S. (1990) *Creating the Second World War*. Pinter, London.

Dalton, R. J. (1988) *Citizen politics in Western Democracies*. Chatham House, Chatham NJ.

—— (1991) 'The dynamics of party system change', in Reif, K. and Inglehart, R. (eds) *Eurobarometer: the dynamics of European public opinion. Essays in honour of Jaquess-Réné Rabier*. Macmillan, London, pp. 18–29.

Davie, G. (2000) *Religion in Europe: a memory mutates*. Oxford University Press, Oxford.

Davies, N. (1996) *Europe: a history*. Oxford University Press, Oxford.

—— (1999) *The Isles: a history*. Macmillan, London.

Dear, M. J. (1988) 'The postmodern challenge: reconstructing human geography', *Transactions of the Institute of British Geographers* NS, 13: 262–74.

Deutsch, K. (1978) *The Analysis of International Relations*. Second edition. Prentice Hall, Englewood Cliffs NJ.

—— (1981) 'The crisis of the state', *Government and Opposition* 16: 331–43.

Dicken, P. (1998) *Global Shift: the transformation of the world economy*. Third edition. Sage, London.

Dodds, K. (1997) *Geopolitics in Antarctica: views from the Southern Ocean rim*. John Wiley, Chichester.

—— (2000) *Geopolitics in a Changing World*. Pearson, Harlow.

Dodds, K. and Sidaway, J. D. (2004) 'Halford Mackinder and the "geographical pivot of history": a centennial retrospective', *Geographical Journal* 170: 292–7.

Doel, M. A. (1993) 'Proverbs for paranoids: writing geography on hollowed ground', *Transactions of the Institute of British Geographers* NS 18: 377–94.

Domesh, M. (1991) 'Towards a feminist historiography of geography', *Transactions of the Institute of British Geographers* NS 16, 95–104.

Donner, M. (1999) 'Endgame in Kosovo', *New York Review of Books* 46: 8, 6 May.

Dos Santos, T. (1973) 'The crisis of development theory and the problem of dependency in Latin America', in Bernstein, H. (ed.) *Underdevelopment and Development*. Penguin, Harmondsworth.

Doyle, R. (2004) 'Energy geopolitics', *Scientific American* October, p. 36.

Driver, F. (1992) 'Geography's empire: histories of geographical knowledge', *Environment and Planning D: Society and Space* 10: 23–40.

Duchacek, I. D. (1973) *Powermaps: comparative politics of constitutions*. ABC-Clio Press, Santa Barbara CA and Oxford.

Ehrlich, P. (1968) *The Population Bomb*. Ballantine Books, New York.

Escobar, A. (1995) *Encountering Development*. Princeton University Press, Princeton NJ.

European Commission (2001) *Working for the Regions*. European Commission, Brussels.

Fleure, H. J. (1919) 'Human regions', *Scottish Geographical Magazine* 35: 94–105.

Francione, G. L. (1996) *Rain without Thunder: the ideology of the animal rights movement.* Temple University Press, Philadelphia PA.

Friedmann, J. and Alonso, W. (1964) *Regional Development and Planning.* MIT Press, Cambridge MA.

Fukuyama, F. (1992) *The End of History and the Last Man.* Free Press, New York.

—— (2004) *State Building: governance and the world order in the 21st century.* Profile Books, London.

Garrity, P. J. (1997) 'How to think about Henry Kissinger', *On Principle* 5(3), John Ashbrook Center for Public Affairs, Ashland University, Ashland OH.

Gellner, E. (1983) *Nations and Nationalism*. Blackwell, Oxford.

Giddens, A. (1985) *The Nation State and Violence*. Polity Press, Oxford.

Glassner, M. I. (1990) *Neptune's Domain: a political geography of the sea.* Unwin Hyman, London.

Glassner, M. I. and Fahrer, C. (2003) *Political Geography.* Third edition. Wiley, New York.

Glenny, M. (1999) *The Balkans 1804–1999: nationalism, war and the great powers.* Granta Books, London.

Godlewska, A. and Smith, N. (eds) (1994) *Geography and Empire*. Blackwell, Oxford.

Gottmann, J. (1952) *La politique des états et leur géographie.* Armand Colin, Paris.

Gregory, D. (1978) *Ideology, Science and Human Geography.* Hutchinson, London.

Gudgin, G. and Taylor, P. J. (1979) *Seats, Votes and the Spatial Organisation of Elections.* Pion, London.

Haraway, D. (1991) *Simians, Cyborgs and Women*. Routledge, London.

Harley, J. B. (1975) *Ordnance Survey Maps: a descriptive manual.* Ordnance Survey, Southampton.

—— (1988) 'Maps, knowledge, and power', in Cosgrove, D. and Daniels, S. (eds) *The Iconography of Landscape*. Cambridge University Press, Cambridge, pp. 277–312.

Hartshorne, R. (1939) *The Nature of Geography: a critical survey of current thought in the light of the past.* Association of American Geographers, Lancaster PA.

—— (1958) 'The concept of geography as a science of space from Kant and Humboldt to Hettner', *Annals of the Association of American Geographers* 48: 97–108.

Harvey, D. (1973) *Social Justice and the City*. Edward Arnold, London.

—— (1979) 'Monument and myth', *Annals of the Association of American Geographers* 69: 362–81.
—— (1982) *The Limits to Capital.* Blackwell, Oxford.
—— (1985) *The Urbanization of Capital.* Blackwell, Oxford.
—— (1989) *The Condition of Postmodernity: an inquiry into the origins of social change*. Blackwell, Oxford.
—— (1996) *Justice, Nature and the Geography of Difference.* Blackwell, Oxford.
—— (2000) *Spaces of Hope*. Edinburgh University Press, Edinburgh.
Hastings, A. (1997) *The Construction of Nationhood, Ethnicity, Religion and Nationalism.* Cambridge University Press, Cambridge.
Haushofer, K. (1931–4) *Macht und Erde.* 3 volumes. B. G. Teubner, Leipzig.
Heffernan, M. (1998) *The Meaning of Europe: geography and geopolitics.* Arnold, London.
Helin, R. A. (1967) 'The volatile adminstrative map of Rumania', *Annals of the Association of American Geographers* 57: 481–502.
Herbertson, A. J. and Herbertson, F. D. (1899) *Man and his Work: an introduction to human geography.* Black, Edinburgh.
Hirst, P. and Thompson, G. (1996) *Globalisation in Question.* Polity Press, Cambridge.
Hobsbawm, E. J. (1990) *Nations and Nationalism since 1780.* Cambridge University Press, Cambridge.
Hodder, R. (1997) *Merchant Princes of the East.* John Wiley, Chichester.
Hoekman, B. and Kostecki, M. (1995) *The Political Economy of the World Trading System: from GATT to WTO.* Oxford University Press, Oxford.
Inman, R. P. and Rubinfeld, D. L. (1998) *Subsidiarity and the European Union.* National Bureau of Economic Research, Washington DC.
Jackson, P. (1989) *Maps and Meaning*. Unwin Hyman, London.
—— (2000) 'Difference', in Johnston, R. J., Gregory, D., Pratt, G., and Watts, M. (eds) *The Dictionary of Human Geography*. Fourth edition. Blackwell, Oxford, pp. 174–5.
Jenks, C. (ed.) (1992) *The Postmodern Reader.* Academy Editions, London.
Jessop, B. (1994) 'Post-Fordism and the state', in Amin, A. (ed.) *Post-Fordism: a reader.* Blackwell, Oxford, pp. 251–79.
—— (1997) 'A neo-Gramscian approach to the regulation of urban regimes: accumulation strategies, hegemonic projects and governance', in Lauria, M. (ed.) *Reconstructing Urban Regime Theory: regulating urban politics in a global economy.* Sage, London, pp. 51–76.
Johnston, R. J. (1979) *Political, Electoral and Spatial Systems*. Clarendon Press, Oxford.
—— (1982) *Geography and the State.* Macmillan, London.
—— (1991) *Anglo-American Human Geography since 1945.* Fifth edition. Edward Arnold, London.

Johnston, R. J., Knight, D. B., and Kofman, E. (eds) (1988a) *Nationalism, Self-determination and Political Geography.* Croom Helm, London.

Johnston, R. J., Pattie, C. J. and Allsop, J. G. (1988b) *A Nation Dividing? The electoral map of Great Britain 1979–1987.* Longman, London.

Jones, M. (1997) 'Spatial selectivity of the state? The regulationist enigma and local struggles over economic governance', *Environment and Planning A* 29: 831–64.

Jones, M. and McLeod, G. (2004) 'Regional spaces, spaces of regionalism: territory, insurgent politics and the English question', *Transactions of the Institute of British Geographers*, NS 29: 433–52.

Jones, M., Jones, R., and Wood, M. (2004) *An Introduction to Political Geography: space, place and politics*. Routledge, London.

Juda, L. (1996) *International Law and Ocean Use Management: the evolution of ocean governance.* Routledge, London.

Kearns, G. P. (1984) 'Closed space and political practice: Frederick Jackson Turner and Halford Mackinder', *Environment and Planning D: Society and Space* 2: 23–34.

—— (1996) 'AIDS and medical geography: embracing the other?', *Progress in Human Geography* 20, 123–31.

Keeble, D. E. (1989) 'Core–periphery disparities, recession and new regional dynamics in the European Community', *Geography* 74: 1–11.

Klingemann, H.-D. and Fuchs, D. (eds) (1995) *Citizens and the State*. Oxford University Press, Oxford.

Knox, P. and Agnew, J. (1998) *The Geography of the World Economy.* Arnold, London.

Kofman, E. and Peake, L. (1990) 'Into the 1990s: a gendered agenda for political geography', *Political Geography Quarterly* 9: 313–36.

Kuhn, T. S. (1970) *The Structure of Scientific Revolutions.* Second edition. University of Chicago Press, Chicago IL.

Kumar, K. (1988) *The Rise of Modern Society: aspects of the social and political development of the West.* Blackwell, Oxford.

Lawson, P. (1993) *East India Company: a history.* Longman, London.

Lee, R. and Wills, J. (1997) *Geographies of Economies*. Arnold, London.

Lefebvre, H. (1992) *The Production of Space.* Blackwell, Oxford.

Leyshon, A. and Thrift, N. J. (1997) *Money/Space: geographies of monetary transformation.* Routledge, London.

Limerick, P. (1987) *Legacy of Conquest: the unbroken past of the American West.* Norton, New York.

Lipson, E. (1956) *The Economic History of England, Vols 2 and 3: the age of mercantilism*. Sixth edition. A. & C. Black, London.

Livingstone, D. N. (1994) 'Climate's moral economy: science, race and place in post-Darwinian British and American geography', in Godlewska, A. and Smith, N. (eds) *Geography and Empire.* Blackwell, Oxford, chapter 7, pp. 132–54.

Lodge, J. (1991) *The Democratic Deficit and the European Parliament*. Fabian Society Discussion Paper No. 4, Fabian Society, London.

—— (1996) *The 1994 Elections to the European Parliament*. Pinter, London.

Lorwin, V. R. (1966) 'Belgium: religion, class, and language in national politics', in Dahl, R. A. (ed.) *Political Oppositions in Western Democracies*. Yale University Press, New Haven CT and London.

Lowe, P. and Goyder, J. (1983) *Environmental Groups in Politics*. Allen & Unwin, London.

MacEwan, A. and M. (1987) *Greenprints for the Countryside*. Allen & Unwin, London.

Mackinder, Sir H. (1904) 'The geographical pivot of history', *Geographical Journal* 23: 421–37.

—— (1919) *Democratic Ideals and Reality: a study of the politics of reconstruction*. Constable, London.

McKinley, C. (1950) 'The Valley Authority and its alternatives', *American Political Science Review* 44: 607–30.

Mann, M. (1984) 'The autonomous power of the state: its origins, mechanisms and results', *Archives européennes de sociologie* 25: 185–213.

Marsh, G. P. (1964 – first edition 1864) *Man and Nature*. Harvard University Press, Cambridge MA.

Martin, G. J. (1980) *The Life and Thought of Isaiah Bowman*. Archon Books, Hamden CT.

Mason, D. (1995) *Race and Ethnicity in Modern Britain*. Oxford University Press, Oxford.

Massey, D. (1979) 'In what sense a regional problem?', *Regional Studies* 13: 233–43.

—— (1984) *The Spatial Division of Labour*. Macmillan, London.

—— (1995) *Space, Place and Gender*. Polity Press, Cambridge.

Meinig, D. W. (1960) 'Commentary on W. P. Webb, "Geographical–historical concepts in American history"', *Annals of the Association of American Geographers* 50: 95–6.

Michalak, W. and Gibb, R. (1997) 'Trading blocs and multilateralism in the world economy', *Annals of the Association of American Geographers* 87: 264–79.

Momsen, J. and Kinnaird, V. (eds) (1993) *Different Places, Different Voices: gender and development in Africa, Asia and Latin America*. Routledge, London.

Monmonier, M. (1991) *How to Lie with Maps*. University of Chicago Press, Chicago IL and London.

Moseley, C. and Asher, R. E. (eds) (1993) *Atlas of the World's Languages*. Routledge, London.

Muehrcke, P. C. (1978) *Map Use: reading analysis and interpretation*. J. P. Publications, Madison WI.

Nairn, T. (1981) *The Break-up of Britain: crisis and neo-nationalism*. Second edition. Verso, London.

Nicholson, H. (1946) *The Congress of Vienna: a study in allied unity*. Constable, London.

Nicholson, P. (1984) 'Aristotle: ideals and realities', in Redhead, B. (ed.) *Political Thought from Plato to NATO*. BBC, London, pp. 30–44.

O'Loughlin, J. and Heske, H. (1991) 'From "Geopolitik" to "géopolitique". Converting a discipline for war to a discipline for peace', in Kliot, N. and Waterman, S. (eds) *The Political Geography of Conflict and Peace*. Belhaven Press, London.

O'Loughlin, J. and van der Wusten, H. (1993) 'The political geography of war and peace 1890–1991', in Taylor, P. (ed.) *Political Geography of the Twentieth Century*. Belhaven Press, London, pp. 63–113.

Olsson, G. (1980) *Birds in Egg/Eggs in Bird*. Pion, London.

—— (1992) *Lines of Power/Limits of Language*. University of Minnesota Press, Minneapolis.

O'Riordan, T. (1981) *Environmentalism*. Pion, London.

—— (1989) 'The challenge for environmentalism', in Peet, R. and Thrift, N. (eds) *New Models in Geography*. Unwin Hyman, London, pp. 77–101.

—— (1996) 'Environmentalism on the move', in Douglas, I., Huggett, R., and Robinson, M. *Companion Encyclopaedia of Geography*. Routledge, London.

Ormeling, F. J. (1983) *Minority Toponyms on Maps*. Department of Geography, University of Utrecht, Utrecht.

Ó Tuathail, G. (1996) *Critical Geopolitics*. University of Minnesota Press, Minneapolis.

—— (1998) 'Postmodern geopolitics? The modern geopolitical imagination and beyond', in Ó Tuathail, G. and Dalby, S. (eds) *Rethinking Geopolitics*. Routledge, London, pp. 16–38.

Ó Tuathail, G. and Dalby, S. (eds) (1996) *Rethinking Geopolitics*. Routledge, London.

Ó Tuathail, G., Dalby, S., and Routledge, P. (eds) (1998) *The Geopolitics Reader*. Routledge, London.

Paasi, A. (1996) *Territories, Boundaries and Consciousness: the changing geographies of the Finnish–Russian border*. Belhaven Press, London.

Painter, J. (1995) *Politics, Geography and 'Political Geography': a critical perspective*. Arnold, London.

Panebianco, A. (1988) *Political Parties: organisation and power*. Cambridge University Press, Cambridge.

Parker, G. (1988) *Geopolitics, Past, Present and Future*. Pinter, London.

Parker, W. H. (1982) *Mackinder: geography as an aid to statecraft*. Clarendon Press, Oxford.

Pauser, F. (1938) *Spaniens Tor zum Mittelmeer*. B. G. Teubner, Leipzig.

Peach, C. (1996) 'Good segregation, bad segregation', *Planning Perspectives* 11: 379–98.

Peach, C., Robinson, V., and Smith, S. J. (eds) (1981) *Ethnic Segregation in Cities.* Croom Helm, London.

Peet, R. (1985) 'The social origins of environmental determinism', *Annals of the Association of American Geographers*, 75: 309–33.

—— (1998) *Modern Geographic Thought.* Blackwell, Oxford and Boston MA.

Penk, A. (1917) *Über politische Grenzen. Rede zum Antritt des Rektorates der Königliche Friedrich-Wilhelms Universität in Berlin.* Königliche Friedrich-Wilhelms Universität, Berlin.

Poetker, J. S. (1967) *The Monroe Doctrine.* Charles E. Merrill, Columbus OH.

Prescott, J. V. R. (1987) *Political Frontiers and Boundaries.* Unwin Hyman, London.

Ratzel, F. (1897) *Politische Geographie*. R. Oldenbourg, München and Berlin.

Redhead, B. (ed.) (1984) *Political Thought from Plato to NATO.* BBC, London.

Rhodes, R. (1997) *Understanding Governance.* Open University Press, Buckingham.

Roberts, P. (2004) *The End of Oil.* Bloomsbury Publishing, London.

Robinson, T. D. (1986) *Stones of Aran: pilgrimage*. Penguin, London.

Rossiter, D. J., Johnston, R. J., and Pattie, C. J. (1999a) *Boundary Commissions: redrawing the UK's map of Parliamentary constituencies.* Manchester University Press, Manchester.

Rossiter, D. J., Johnston, R. J., Pattie, C. J., Dorling, D. F., Turnstall, H., and McAllister, I. (1999b) 'Changing biases in the operation of the UK's electoral system, 1950–1997', *British Journal of Politics and International Relations* 1.

Rostow, W. W. (1971) *The Stages of Economic Growth: a non-communist manifesto.* Cambridge University Press, Cambridge.

Rowe, C. (1984) 'Plato; the search for an ideal form of state', in Redhead, B. (ed.) *Political Thought from Plato to NATO*. BBC, London, pp. 18–29.

Royal Commission on Legal Services (1979) *Final Report*. Cmnd 7648, HMSO, London.

Runte, A. (1979) *National parks: the American Experience.* University of Nebraska Press, Lincoln.

Sachs, W. (1992) *The Development Dictionary. A guide to knowledge as power.* Zed Books, London.

Sack, R. D. (1980) *Conceptions of Space in Social Thought. A geographic perspective.* Macmillan Press, London.

—— (1986) *Human Territoriality: its theory and history.* Cambridge University Press, Cambridge.

Said, E. (1995) *Orientalism.* Second edition. Penguin, London.

Saiko, T. (2001) *Environmental Crises.* Prentice Hall, Harlow.

Sander, G. and Rossler, M. (1994) 'Geography and empire in Germany', in Godlewska, A. and Smith, N. (eds) *Geography and Empire.* Basil Blackwell, Oxford, pp. 115–29.

Scarman, The Right Honourable the Lord L. (1981) *The Brixton Disorders, 10–12 April 1981.* Cmnd 8427, HMSO, London.

Schram Stokke, O. and Vidas, D. (1996) *Governing the Antarctic.* Cambridge University Press, Cambridge.

Scudder, E. S. (1939) *The Monroe Doctrine and War and Peace.* Nelson, London.

Sheail, J. (1981) *Rural Conservation in Inter-war Britain.* Clarendon Press, Oxford.

Sibley, D. (1995) *Geographies of Exclusion: society and difference in the west.* Routledge, London.

Sinnhuber, K. (1954) 'Central Europe – Mitteleuropa – l'europe centrale: an analysis of a geographical term', *Transactions and Papers, Institute of British Geographers* 20: 15–35.

—— (1964) 'The representation of disputed political boundaries in general atlases', *Cartographic Journal* 1: 20–8.

Skelton, R. A. (1962) 'The origins of the Ordnance Survey of Great Britain', *Geographical Journal* 128: 415–26.

Smith, D. M. (1994) *Geography and Social Justice.* Blackwell, Oxford.

Smith, N. (1984) 'Isaiah Bowman: political geography and geopolitics', *Political Geography Quarterly* 3: 69–76.

—— (2003) *American Empire: Roosevelt's geographer and the prelude to globalisation.* University of California Press, Berkeley.

Smith, S. J. (1989) *The Politics of 'Race' and Residence: citizenship, segregation and white supremacy in Britain.* Polity Press, Cambridge.

Soja, E. W. (1989) *Postmodern Geographies.* Verso, London and New York.

Spykman, N. (1938) 'Geography and foreign policy', *American Political Science Review* 32: 28–50 and 213–36.

—— (1942) 'Frontiers, security and international relations', *Geographical Review* 32: 436–47.

Stamp, Sir D. (1966) 'Philatelic cartography', *Geography* 51: 179–97.

Stoddart, D. R. (1966) 'Darwin's impact on geography', *Annals of the Association of American Geographers* 56: 683–98.

—— (1981) *Geography, Ideology and Social Concern.* Blackwell, Oxford.

Suter, K. (1991) *Antarctica: private property or public heritage?* Pluto Press, Australia.

Taaffe, P. (1995) *The Rise of Militant: Militant's thirty years 1964–1994.* The Socialist Party, London.

Taylor, P. J. (1985) *Political Geography: world-economy, nation-state and locality.* First edition. Longman, Harlow.

—— (1993) *Political Geography of the Twentieth Century: a global analysis.* Belhaven Press, London.

—— (2000) 'Critical geopolitics', in Johnston, R. J., Gregory, D., Pratt, G., and Watts, M. (eds) *The Dictionary of Human Geography.* Fourth edition. Blackwell, Oxford, pp. 125–6.

Taylor, P. J. and Johnston, R. J. (1979) *Geography of Elections*. Penguin, London.

Taylor, P. J. and Flint, C. (1999) *Political Geography: world-economy, nation-state and locality*. Fourth edition. Longman, London.

Thrasher, M. and Rallings, C. (eds) (2000) *British Electoral Facts 1832–1999*. Sixth edition. Ashgate, Aldershot.

Thrift, N. J. (1995) 'A hyperactive world', in Johnston, R. J., Taylor, P. J., and Watts, M. J. (eds) *Geographies of Global Change: remapping the world in the late twentieth century*. Blackwell, Oxford, pp. 18–35.

Thrower, N. J. W. (1999) *Maps and Civilization: cartography in culture and society*. University of Chicago Press, Chicago IL and London.

Tickle, A. and Welsh, I. (eds) (1998) *Environment and Society in Transition: Central and Eastern Europe*. Longman, London.

Turner, F. J. (1894) *The Significance of the Frontier in American History*. Annual Report of the American Historical Association, US Government Printing Office, Washington DC.

Tyner, J. A. (2004) 'Territoriality, social justice and gendered revolutions in the speeches of Malcolm X', *Transactions of the Institute of British Geographers* NS 29: 330–43.

Unwin, T. (1998) *A European Geography*. Longman, Harlow.

Unwin, T. and Hewitt, V. (2001) 'Banknotes and national identity in central and eastern Europe', *Political Geography* 20: 1005–28.

Urry, J. (1981) *The Anatomy of Capitalist Societies: the economy, civil society and the state*. Macmillan, London.

Vujakovic, P. (1989) 'Mapping for world development', *Geography* 74: 97–105.

Wainwright, H. (2004) *Reclaim the State: experiments in popular democracy*. Verso, London.

Waldron, J. (1990) *The Right to Private Property*. Clarendon, Oxford.

Wallerstein, I. (1974) *The Modern World System: capitalist agriculture and the origins of the European world-economy in the sixteenth century*. Academic Press, New York.

—— (1979) *The Capitalist World Economy*. Cambridge University Press, Cambridge.

—— (1980) *The Modern World System II: Mercantilism and the consolidation of the European world economy 1600–1750*. Academic Press, New York.

—— (1983) *Historical Capitalism*. Verso, London.

—— (1984) *The Politics of the World Economy*. Cambridge University Press, Cambridge.

Wanklyn, H. (1961) *Friedrich Ratzel: biographical memoir and bibliography*. Cambridge University Press, Cambridge.

Waters, M. (1995) *Globalisation*. Routledge, London.

Watts, M. (2000) 'Development', in Johnston, R. J., Gregory, D., Pratt, G., and Watts, M. (eds) *The Dictionary of Human Geography*. Fourth edition, Blackwell, Oxford, pp. 166–71.

Westminster City Council (2002) *Evaluation of Accuracy and Reliability of 2001 Census.* Westminster City Council, London.

Wheen, F. (1999) *Karl Marx.* Fourth Estate, London.

Whittlesey, D. (1939) *Earth and State: a study in political geography.* Henry Holt, New York.

Wild, A. (2000) *The East India Company: trade and conquest from 1600.* Lyons Press, New York.

Wilkinson, H. R. (1951) *Maps and Politics: a review of the ethnographic cartography of Macedonia.* Liverpool University Press, Liverpool.

Williams, A. M. (1994) *The European Community: the contradictions of integration.* Second edition. Blackwell, Oxford.

Wise, M. (1984) *The Common Fisheries Policy of the European Community.* Methuen, London.

Wise, M. and Gibb, R. (1993) *Single Market to Social Europe.* Longman, London.

Wood, D. (1993) *The Power of Maps.* Routledge, London.

Wood, W. B. (2001) 'Geographic aspects of genocide: a comparison of Bosnia and Rwanda', *Transactions of the Institute of British Geographers* NS 26: 57–76.

Woods, M. (1998) 'Rethinking elites: networks, space and local politics', *Environment and Planning A* 30: 2101–19.

Young, I. M. (1990) *Justice and the Politics of Difference.* University of Princeton Press, Princeton NJ.

Zelinsky, W. and Williams, C. H. (1988) 'The mapping of language in North America and the British Isles', *Progress in Human Geography* 12: 337–68.

Index